An Introduction to Population Genetics

THEORY AND APPLICATIONS

An Introduction to Population Genetics

THEORY AND APPLICATIONS

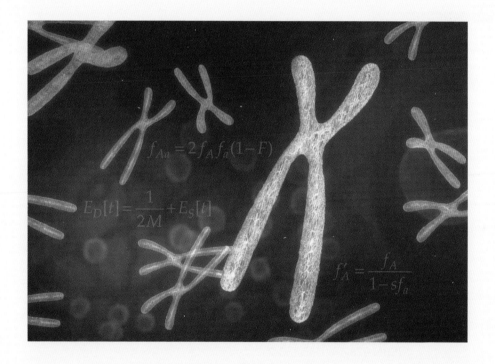

Rasmus Nielsen • Montgomery Slatkin

Sinauer Associates, Inc. Publishers
Sunderland, Massachusetts U.S.A.

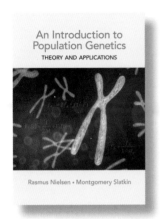

About the cover
The cover illustrates the range of topics presented in the book, which emphasizes both the biological and theoretical aspects of population genetics.

An Introduction to Population Genetics: Theory and Applications
Copyright © 2013 by Sinauer Associates, Inc.

For information:
Sinauer Associates, P.O. Box 407, Sunderland, MA 01375 U.S.A.
Fax: 413-549-1118
Email: publish@sinauer.com
Internet: www.sinauer.com

Library of Congress Cataloging-in-Publication Data

Nielsen, Rasmus, 1970-
An introduction to population genetics : theory and applications / Rasmus Nielsen, Montgomery Slatkin.
      p. ; cm.
Includes index.
ISBN 978-1-60535-153-7
I. Slatkin, Montgomery. II. Title.
[DNLM: 1. Genetics, Population. 2. Gene Frequency--genetics. 3. Genetic Drift. 4. Models, Genetic. QU 450]

576.5'8--dc23
                                    2012046169

Printed in China

6 5 4 3 2 1

Brief Contents

Contents

Preface

This book was born out of our belief that coalescence theory provides an easy and intuitive way to understand complex population genetic problems. We have tried to combine coalescence theory with classical population genetics and present applications of the theory to human and other populations. Early versions of this book have been used for a one-semester undergraduate course in population genetics at U.C. Berkeley. The book is intended for undergraduate and graduate students who have some basic knowledge of biology and genetics and who are not afraid of quantitative thinking. The theory we present requires only basic algebra. We introduce some ideas from probability and statistics that are an integral part of modern population genetics.

We begin with the basic definitions in population genetics, leading up to the concept of Hardy–Weinberg frequencies (Chapter 1). Then we describe the way allele frequencies change when there is no selection—genetic drift (Chapter 2). In Chapters 3 and 4, we introduce the coalescent approach to understanding genetic drift and describe how that leads directly to the analysis of data within and between populations. We apply this theory to data in Chapter 5, showing how the history of populations can be inferred. The theory of two loci (linkage disequilibrium) is presented and then used for identifying genes that affect inherited diseases in humans. The next several chapters introduce the theory of natural selection. Chapter 7 describes selection acting alone, and Chapter 8 shows how we can combine estimates of selection with estimates of genetic drift to predict patterns of change at the DNA and amino acid sequence level. Chapter 9 presents several ways that population geneticists test for the action of natural selection. Chapter 10 offers examples of more complex kinds of selection that result from interactions among individuals in a population and among inherited elements within the genome. Finally, we present the elements of quantitative genetics—the study of phenotypes affected by multiple genetic loci.

We thank the students in our course for allowing us to use them as guinea pigs for this experiment. We also thank members of the Nielsen and Slatkin labs at U.C. Berkeley for comments on the manuscript, in particular Mike DeGiorgio, Kelley Harris, Mason Liang, and Vitor Aguiar. Finally, we would like to thank our editor, Andy Sinauer, for his encouragement and the opportunity to publish this work; our production editor, Martha Lorantos; and the entire production staff at Sinauer Associates for their excellent guidance in transforming our scribbles into a cohesive book.

An Introduction to Population Genetics

THEORY AND APPLICATIONS

Introduction

IN THIS BOOK, we will introduce the principles of population genetics. These principles are applicable to all genetic variants that can be distinguished by some means and that can be transmitted from parents to offspring. We will call any variants with these properties *alleles*. For example, at a particular nucleotide position in the *MC1R* gene, the human genome may have either C or T. The C and the T alleles can be distinguished by sequencing the *MC1R* gene. Individuals with T at this position—position 478— will have red hair and freckles. Most of the time, alleles obey Mendel's First Law, in which case we call them *Mendelian alleles*.

Population genetics is the study of alleles in populations. The subject is both predictive—predicting the future composition of a population from its current composition—and retrospective—understanding what determined the current composition of a population.

The predictive aspect of population genetics, as it was developed in the twentieth century, deals with changes in the frequencies of specific alleles and genotypes under various conditions. The more recently developed retrospective approach to population genetics focuses on the ancestry of genes, and is called *coalescent theory*. Although looking backward in time rather than forward takes some getting used to, it is often the simplest way to understand the history of populations, particularly when DNA sequence data is available for analysis. We will present both approaches to population genetics. They are equivalent in many ways, and both are necessary to understand the explosive growth of population genetics that has happened in response to new DNA sequencing methods.

Types of Genetic Data

The C/T alternative at position 478 in *MC1R* is an example of a **single nucleotide polymorphism** (**SNP**; pronounced "snip"). This is one of a few kinds of genetic variants commonly used in population genetics studies.

Individuals homozygous for the T allele in position 478 of the *MC1R* gene tend to have freckles and red hair. *MC1R* codes for a protein called the melanocortin 1 receptor. This receptor transmits signals relating to the production of melanin (pigment) in skin cells. The mutation in position 478 disrupts the protein and causes an increase in the production of the red/yellow pigment phaeomelanin instead of the brown/black pigment eumelanin.

Another type of variant is the insertion or deletion of a few nucleotides, called an indel. One example of an indel variant, is CFTR-ΔF508 which causes cystic fibrosis. The CFTR gene codes for a long transmembrane protein involved in governing the osmotic balance of cells. The variant ΔF508 has a three-base deletion in the coding sequence that results in the absence of the 508th amino acid, phenylalanine (denoted by *F* in standard biochemical notation). This variant is not a SNP, but it is genetically transmittable from parents to offspring, and hence is an allele. People homozygous for ΔF508 have cystic fibrosis, a disorder attributable in part to poor regulation of osmotic balance. The frequency of ΔF508 is about 2% in European populations, and much smaller in other populations.

Another kind of genetic variant is created by the tendency of the DNA replication machinery to miscopy repeated sequences in the genome. For example, suppose the sequence ATGGCTGCACACACACACACATGCTGA appears on one chromosome sampled from a population. On this individual's chromosome, the CA motif is repeated seven times, written as $(CA)_7$. Another individual's chromosome may have six, eight, or some other number of repeats, $(CA)_n$, at this position. The characteristic number of repeats is transmitted during meiosis, with a small possibility of error, and so the variants are considered to be alleles. Variants of this type are called **simple sequence repeats (SSRs)** or, more commonly, **microsatellites**. There are many microsatellite loci in humans and other vertebrates; in the human genome, there is roughly one CA microsatellite for every 6000 bases. Microsatellites are very useful in the study of human population, because each one is likely to have several alleles that differ in the repeat number.

Detecting Differences in Genotype

The ultimate way to detect differences in genotype is to obtain the DNA sequence of each locus of interest, but even with the rapid development of sequencing methods, that is not feasible most of the time. It is too expensive and, for species other than a few model organisms such as *Drosophila melanogaster* and *Mus musculus*, too difficult. When sequencing cannot be done, population geneticists use other information that reveals something, if not everything, about genetic differences among individuals.

Until the 1960s, visible differences in phenotype, such as banding patterns in snails and colors in flowers that conformed to Mendel's First Law, provided the only data for population geneticists to analyze. This was the era of **ecological genetics**. In the 1960s, the first biochemical methods for detecting differences in DNA sequence became widely used. **Protein electrophoresis** was used to characterize alleles according to the speed a stained protein moved on a gel under standard conditions. For example, the ADH gene in *D. melanogaster* was found to have two alleles, denoted by *F* (fast) and *S* (slow). Protein electrophoresis could reveal some, but not all, changes in a DNA sequence. Only those changes in amino acid sequence that resulted in differences in mobility could be detected. Nevertheless, protein electrophoresis created a revolution in population genetics. For the first time, abundant data became available for a wide variety of species. Furthermore, the data closely reflected differences in the DNA sequence of single genes. Protein electrophoresis was used to systematically survey most kinds of plants, animals, and microorganisms, and population genetics theory developed rapidly in response.

In the 1980s, the **polymerase chain reaction** (**PCR**), together with Sanger sequencing, provided access to DNA sequencing. In humans and other animals, the technology was often used to sequence DNA sequences of **mitochondrial DNA** (**mtDNA**). The mitochondrion carries its own DNA, a circular molecule. Mitochondrial DNA was used then, and is still used today, as a population genetic marker because mtDNA mutates much faster than nuclear DNA, and segments of the molecule are, therefore, highly variable. In addition, since each cell carries many mitochondria, sequencing of the molecule is easy, because mtDNA is highly abundant in the cell.

The targeted sequencing of mtDNA quickly became one of the major methodologies in population genetic analyses. However, it provided information about the mitochondria only—not about genome-wide processes. Restriction enzymes, originally discovered in the 1970, provided a way to detect differences in DNA sequence at a genome-wide scale. A given restriction enzyme cuts DNA whenever a particular sequence is encountered. For example, *eco*R1 cuts a chromosome whenever the sequence GAATTC, the recognition sequence for *eco*R1, is found. Other restriction enzymes recognize and cut other sequences. After prepared chromosomes are exposed to a particular restriction enzyme, the DNA is cut into a number of fragments whose lengths depend on the locations of the recognition sites. The sizes of the resulting fragments can be determined using gel electrophoresis, since the distance each fragment moves on a gel depends on its length. By careful analysis, using multiple restriction enzymes, it is possible to find the genomic location of restriction sites for each enzyme.

Once this had been done for a few model organisms, including humans, it was possible to detect differences in the sequence of each restriction site, because even a single change in the recognition sequence causes the restriction enzyme to not cut the DNA at that location. For example, if a chromosome had GAGTTC instead of GAATTC, *eco*R1 would not cut the DNA there. The absence of the restriction site would be detectable from differences in fragment lengths seen on a gel. Although surveying genetic variation using restriction enzymes was both time-consuming and expensive, it was a breakthrough for both population and human genetics. For the first time, differences in the sequence of noncoding as well coding DNA could be detected without having to sequence each individual chromosome, a very laborious process at the time. The whole genomes of model organisms became available for population genetic analysis, not only the small fractions of these genomes that code for proteins.

The next major advance in surveying genetic variation came in the early 1990s, with the development of efficient methods for genotyping microsatellite loci. Recall that alleles at a microsatellite locus differ in the number of repeats of a DNA motif that is usually two to six bases in length. On either side of a microsatellite locus is nonrepetitive DNA. Once a microsatellite allele has been found, primers for a polymerase chain reaction (PCR) that bind uniquely to the flanking nonrepetitive DNA can be designed. Once that is done, it is relatively easy to determine the length of the fragment between the PCR primers without sequencing that fragment. The length of the fragment indicates the number of repeat units.

More recently, population genetics is undergoing another revolution. New sequencing methods, called **next-generation-** or **new generation sequencing** (**NGS**), allows for cheap direct sequencing of multiple genomes. The 1000 Genomes Project has already sequenced the genomes of thousands of humans from populations throughout the world. Similar projects are being carried out for *Drosophila*, mice, *Arabidopsis*, and many domesticated plants and animals for which complete genomic sequences are available. NGS is also used in the study of natural populations; it is facilitated by the availability of methods for extracting subsets of a genome—for example, all the protein-coding sequences, or a random subset of sequences—a less expensive process than complete sequencing. Using these technologies, population genetic analyses are now finally based on directly sequenced DNA from large parts of the genomes of many species. The resulting dramatic increase in the amount of data available has created unprecedented opportunities for population genetic analyses and has led to another period of rapid growth in the theory.

1 Allele Frequencies, Genotype Frequencies, and Hardy–Weinberg Equilibrium

MOST READERS OF THIS BOOK will be familiar with the terminology of genetics. But since some terms are defined slightly differently in population genetics than in other areas of genetics and molecular biology, some definitions might be useful at the outset. A locus (plural: *loci*) is a position in the genome where there might be one or more alleles segregating. Some geneticists use the word *locus* as synonymous to *coding gene*. However, in population genetics, the word *locus* is generally used to represent *any position in the genome*. It could be a coding gene, such as the *MC1R* gene; it could be a microsatellite; or it could be a single nucleotide position in the genome, such as position 8,789,654 of chromosome 1 of the human genome. In general, any unit in the genome with one or more alleles is a locus. A **genotype** is the combination of alleles carried by an individual in a particular locus. For example, if an individual is homozygous TT in position 8,789,654 of chromosome 1 of the human genome, then we say that this individual has genotype TT at that locus. A diploid species, such as humans, has two copies of all its chromosomes. For a collection of N diploid individuals, there are $2N$ gene copies at each locus, and there could be one or more alleles.

A major objective of classical population genetics is to understand how allele frequencies change through time. To simplify the analyses of allele frequencies, we often use models where there are two alleles—say, allele A and allele a. We call such models **di-allelic models**. The two alleles could, for example, represent the normal and the red-hair ver-

[handwritten margin note: location w/i genome with 1 or more alleles]

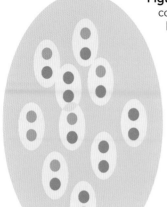

Figure 1.1 A hypothetical population with $N = 10$ individuals, 20 gene copies, and a total of 7 copies of allele A (green) and 13 copies of allele a (red), i.e., $f_A = 7/20$ and $f_a = 13/20$. The genotype frequencies are $f_{AA} = 1/10$, $f_{Aa} = 5/10$, and $f_{aa} = 4/10$.

sion of the *MC1R* gene discussed in the introduction— or two different versions of any other gene. Di-allelic models can also be used to model DNA sequences. At any position in the genome, there are four possible nucleotides, A, C, T, and G, but because mutations are rare in most organisms, you will typically tend to see at most two nucleotides in any particular position in the individuals of the population. For example, in nucleotide position 478 of the *MC1R* gene in humans, most individuals have a C, but some individuals have a T; A and G have not been observed in this position. So we can, at least as a first approximation, use a di-allelic model to describe this position in the genome.

We sometimes depict a population as in **Figure 1.1**. The blue oval represents the population, and the tan ovals within it represent individuals. The red and green balls within the individual ovals represent two alleles segregating in the population, alleles A and a. The combination of alleles within each tan oval represents the genotype of an individual; thus, an oval with a green and a red ball represents a heterozygous individual of genotype Aa.

Allele Frequencies

The frequency of an allele is defined as the number of copies of the allele in the population divided by the total number of gene copies in the population. In a diploid population (in which all individuals carry two copies of each chromosome) with N individuals, there are $2N$ gene copies. So the frequencies of alleles A and a are:

$$f_A = \frac{N_A}{2N} \text{ and } f_a = \frac{N_a}{2N} \tag{1.1}$$

where N_A and N_a are the numbers of A and a alleles segregating in the population, respectively. Of course, the allele frequencies must add up to 1, so $f_A + f_a = 1$. Much population genetic theory concentrates on describing the changes of f_A and f_a with time. If we can describe how we expect allele frequencies to change through time, we have learned a great deal about evolution.

Genotype Frequencies

The allele frequencies in the population can be calculated from the genotype frequencies. In a di-allelic locus, there are three possible genotypes: AA, Aa, and *aa*. If the number of copies of genotypes *AA*, *Aa*, and *aa* are N_{AA}, N_{Aa}, and N_{aa}, respectively, then the genotype frequencies are:

$$f_{AA} = \frac{N_{AA}}{N} \qquad f_{Aa} = \frac{N_{Aa}}{N} \qquad f_{aa} = \frac{N_{aa}}{N} \qquad (1.2)$$

Notice that while the denominator in Equation 1.2 is *N*, the denominator in Equation 1.1 is 2*N*, as there are 2*N* gene copies in a diploid population of *N* individuals. The genotype frequencies will add up to 1: $f_{aa} + f_{Aa} + f_{AA} = 1$.

Individuals of genotype *AA* carry two copies of allele *A* and individuals of genotype *Aa* carry one copy of allele *A*. The allele frequency of allele *A* can, therefore, be calculated as:

$$f_A = \frac{2N_{AA} + N_{Aa}}{2N} = f_{AA} + f_{Aa} / 2 \qquad (1.3)$$

Similarly, $f_a = f_{aa} + f_{Aa}/2$. The proportion of individuals that are heterozygous in the population (f_{Aa}) is called the **heterozygosity** of the population. The proportion that is homozygous ($1 - f_{Aa} = f_{AA} + f_{aa}$), is the **homozygosity** of the population.

K-allelic Loci

A locus in which there are *k* different alleles, where *k* could be any positive natural number, is usually referred to as a ***k*-allelic locus**. Microsatellite loci often have more than two alleles. We can find expressions for allele and genotype frequencies for a general *k*-allelic locus similar to the ones we have already found for a di-allelic locus. For an allele, $i \in \{1, 2,..., k\}$, with N_i copies in the population, the allele frequency is $f_i = N_i/2N$, and for a genotype *ij* (= *ji*), the genotype frequency is $f_{ij} = N_{ij}/N$. The allele frequency can then be calculated from the genotype frequencies as:

$$f_i = f_{ii} + \sum_{j:j\neq i} f_{ij} / 2 \qquad (1.4)$$

The concepts of homozygosity and heterozygosity can also be extended to *k*-allelic loci, with $\sum_i f_{ii}$ being the homozygosity and $\sum_{(i,j):i<j} f_{ij}$ being the heterozygosity. In this book we will mostly concentrate on di-allelic loci, because the mathematical notation is simpler for such loci. However, much of the theory discussed easily extends to loci with more than two alleles.

Example: The MC1R Gene

Let us again consider position 478 of the *MC1R* gene. Suppose we obtain a random sample of 30 individuals from the United States and find 25 individuals of genotype CC, 5 individuals of genotype CT, and 0 individuals of genotype TT. The genotype frequencies can then be estimated as f_{CC} = 25/30 = 0.833; f_{CT} = 5/30 = 0.167; and f_{TT} = 0/30 = 0. The allele frequencies can be estimated as f_C = 0.833 + 0.167/2 = 0.917 and f_T = 1 − 0.917 = 0.083.

Notice here that we used the word *estimated*. We cannot know the true genotype or allele frequencies in the entire population without examining all the individuals in the population, but we can hope that this sample of 30 individuals is representative. Had we taken another sample of 30 different individuals, we might have obtained a slightly different answer.

Hardy–Weinberg Equilibrium

We have seen how allele frequencies can be calculated from genotype frequencies. But can we also predict genotype frequencies from allele frequencies? For example, knowing that the frequency of T in position 478 of the *MC1R* locus is approximately 0.08, what proportion of the population would we expect to have genotype TT?

We can answer this question, but only if we make some assumptions. One particularly useful simplifying assumption is that mating is random, i.e., that individuals mate with each other without regard to genotype. Imagine a pool of parental males and a pool of parental females that mate randomly, i.e, the next generation is produced by randomly choosing the father and the mother from these pools of potential parents independently of each other for each individual in the offspring generation. For now, assume that the allele frequency among males is the same as among females, and that there are only two alleles, A and a, for the locus under consideration. Given these assumptions, the chance that an individual offspring is of genotype AA is given by the probability of receiving an A allele from the father and an A allele from the mother. The probability that an A allele is transmitted to the next generation is simply the frequency of the allele, f_A, because all gene copies have the same probability of transmission under Mendel's First Law. The assumption of random mating ensures that we can multiply the probabilities from the father (f_A) and the mother (f_A), so the probability that an individual in the population is of type of AA is simply f_A^2.

Likewise, an individual offspring can be heterozygous by getting an A allele from the father and an a allele from the mother—or by getting an a allele from the father and an A allele from the mother. The probability than an individual is of genotype Aa is then $f_A f_a + f_a f_A = 2f_A f_a$. Finally, using the same logic, we find that the probability that an individual is homozygous, aa, is f_a^2. The expected proportion of individuals of a particular genotype

TABLE 1.1 Genotype frequencies under Hardy–Weinberg Equilibrium

Genotype	AA	Aa	aa
Frequency	f_A^2	$2f_A f_a$	f_a^2

in the population is simply the genotype probabilities we have calculated, and we have arrived at the famous Hardy–Weinberg equilibrium theory: The **expected homozygosity** in the population is then $f_a^2 + f_A^2$ and the **expected heterozygosity** is $2f_A f_a$.

The reader may previously have encountered Hardy–Weinberg Equilibrium (HWE) theory using the notation p^2, $2pq$, and q^2 for the three genotype probabilities, respectively. Notice that this result is exactly the same as that stated in **Table 1.1**, with f_A replaced by p and f_a replaced by q. We use our notation because it generalizes more easily. As required, the genotype frequencies under HWE will add up to 1:

$$f_A^2 + 2f_A f_a + f_a^2 = (f_A + f_a)^2 = 1 \qquad (1.5)$$

The concept of *probability* used here to derive HWE is discussed in **Box 1.1**. Box 1.1 also discusses the concept of *independence*. The reader may notice that the assumption of random mating implies that we draw alleles independently from male and female parents, allowing us to multiply the allele frequencies together in the offspring population. In terms of the notation from Box 1.1, we could write:

$$
\begin{aligned}
&\Pr\,(\textit{offspring genotype} = AA) \\
&= \Pr\,(\textit{paternal allele} = A) \times \Pr\,(\textit{maternal allele} = A) \qquad (1.6) \\
&= f_A f_A = f_A^2
\end{aligned}
$$

While the basic ideas in Box 1.1 are not prerequisite to an understanding of HWE, they will be used throughout this book, and should be reviewed at this point if they are not already familiar.

An alternative derivation of HWE, based on enumerating all possible matings, is shown in **Box 1.2**. We obtain the same result using that approach, demonstrating that random mating is, in fact, equivalent to independent sampling of paternal and maternal alleles.

Finally, notice that random mating in itself does not change the allele frequencies. The frequency of allele A in the next generation (f_A')

$$f_A' = f_A^2 + 2f_A f_a/2 = f_A^2 + 2f_A(1 - f_A)/2 = f_A \qquad (1.7)$$

will be the same as in the previous generation.

allele freq. remain constant from to generation.

The MC1R *Gene Revisited*

Now let's revisit the question regarding prediction of genotype frequencies in position 478 of the *MC1R* locus. With an allele frequency of 0.08 of allele T in the US population, how many TT homozygotes might we expect? Using HWE theory we will expect the proportions of individuals with genotypes CC, CT, and TT to be $0.92^2 = 0.8464$, $2 \times 0.92 \times 0.08 = 0.1472$, and $0.08^2 = 0.0064$, respectively. Part of the interest in this gene is caused by the fact that individuals with the TT genotype will likely have red hair (Introductory Figure). However, a much larger proportion of the population has red hair

BOX 1.1   Probability and Independence

Although there are different schools of thought regarding definitions of **probability**, we will here think of probability as expressing belief in future events, or outcomes of an experiment. For example, if we toss a coin and make the statement, "The probability of observing a head is ½ and the probability of observing a tail is ½," then we believe heads and tails are equally likely to occur in the next toss. Let X be a variable that indicates the outcome of the coin toss. The variable X can take two different values: H for heads and T for tails. We can then write

$$\Pr(X = H) = ½$$

where $X = H$ denotes the event that the coin toss results in a H. A variable such as X, that can take on different values with different probabilities, is called a **random variable**. In words, we can read the equation above as: "The probability that the random variable X takes on the value H equals one-half," a mathematical way of saying that we think heads and tails are equally likely outcomes of the coin toss.

The **sample space** of a random variable is the set of possible values that the random variable can take on. In the coin-toss case, the sample space is $\{H, T\}$.

Two random variables are **independent** if the outcome of one variable does not affect (our belief in) the outcome of the other variable. For example, if we toss a coin twice, it is reasonable to assume that the result from the first coin toss does not affect the second coin toss, so the two coin tosses are independent of each other. Just because the first coin toss resulted in an H does not mean that we think the next coin toss also will result in an H—as long as the coin is not biased.

If two random variables are independent, we can multiply their probabilities. In the coin toss example, if we let X be the result of the first coin toss, and Y be the result of the second coin toss, we find that the **joint probability** is:

$$\Pr(X = H \text{ and } Y = H) = \Pr(X = H) \times \Pr(Y = H) = 0.5 \times 0.5 = 0.25$$

However, imagine a bag full of fake coins that are biased, half of which give H with probability 0.9 and half of which give T with probability 0.9. If we randomly pick a coin from this bag and toss it twice, these two coin tosses are correlated (not independent). The chance that the first coin toss gives H is still 0.5, because half of the coins in the bag are biased toward H and half are biased toward T. However, if the first coin toss gives H, it is likely that we have picked an H-biased coin, and our belief that the second coin toss will also result in H has increased. The two coin tosses are not, mathematically speaking, independent, and the joint probability of observing an H on both the first and the second coin toss is no longer 0.25. We can no longer obtain the joint probability from the two coin tosses by multiplying the probabilities from each coin toss. These concepts are expanded upon in Appendix A.

BOX 1.2   Derivation of HWE Genotype Frequencies

In the text, we derived the Hardy–Weinberg genotype frequencies by assuming that gametes inherited from the mother and father assorted independently. We derive the frequencies here by considering all possible matings, show in the table below. The frequency of each type of mating is the product of the genotype frequencies. That is what is meant by **random mating**. The genotypes of the offspring then follow from Mendel's First Law (random gamete assortment).

| | | | Offspring | | |
Mother	Father	Frequency	AA	Aa	aa
AA	AA	f_{AA}^2	1	0	0
AA	Aa	$f_{AA}f_{Aa}$	½	½	0
AA	aa	$f_{AA}f_{aa}$	0	1	0
Aa	AA	$f_{Aa}f_{AA}$	½	½	0
Aa	Aa	f_{Aa}^2	¼	½	¼
Aa	aa	$f_{Aa}f_{aa}$	0	½	½
aa	AA	$f_{aa}f_{AA}$	0	1	0
aa	Aa	$f_{aa}f_{Aa}$	0	½	½
aa	aa	f_{aa}^2	0	0	1

The genotype frequencies in the offspring are found by adding over all the families:

$$f'_{AA} = (1)f_{AA}^2 + (½)f_{AA}f_{Aa} + (½)f_{Aa}f_{AA} + (¼)f_{Aa}^2 = (f_{AA} + (1 - f_{Aa}/2)^2 = f_A^2$$

$$f'_{Aa} = (½)f_{AA}f_{Aa} + (1)f_{AA}f_{aa} + (½)f_{Aa}f_{AA} + (½)f_{Aa}^2 + (½)f_{Aa}f_{aa} + (1)f_{aa}f_{AA} + (½)f_{aa}f_{Aa}$$

$$= 2(f_{AA} + f_{Aa}/2)(f_{aa} + f_{Aa}/2) = 2f_A f_a$$

$$f'_{aa} = (¼)f_{Aa}^2 + (½)f_{Aa}f_{aa} + (½)f_{aa}f_{Aa} + (1)f_{aa}^2 = (f_{aa} + f_{Aa}/2)^2 = f_a^2$$

where the prime (′) indicates the frequencies among the offspring. No matter what the genotype frequencies are, one generation of random mating will establish the HWE genotype frequencies.

than the expected 0.64% from this calculation, telling us that other factors are important for the development of red hair than being homozygous TT at position 478 of the *MC1R* locus.

Tay–Sachs Disease

HWE has many applications, including analysis of allele frequencies that impact health in humans: the frequency of individuals affected by diseases caused by recessive deleterious mutations can be predicted from the allele frequencies. An example is Tay–Sachs disease, which causes deterioration of mental and physical abilities and usually ends in death by the age of four. Individuals homozygous for certain mutations in the *HEXA* gene will be affected by this disease. A four-base-pair insertion in the gene, causing a change in reading frame that essentially destroys the function of the gene, is common among Ashkenazi Jews. In fact, the allele frequency of this mutation among Ashkenazi Jews is as high as 2%. What is the proportion of offspring of Ashkenazi Jewish couples that will be affected by Tay–Sachs disease because they are homozygous for the disease mutation? Using HWE, we find the answer to be $0.02^2 = 0.0004$ or 0.04%. This disease risk is sufficiently high that Ashkenazi Jewish couples in the United States and Israel are often genetically screened for Tay–Sachs Disease.

Extensions and Generalizations of HWE

HWE shows that if the allele frequencies are identical in males and females, after one round of random mating, the genotype frequencies can be obtained simply by multiplying together the appropriate allele frequencies. If the allele frequencies are different in males and females, it takes two generations before HWE is established. After one generation of random mating, the allele frequencies in males and females will become the same. The next generation of random mating then establishes HWE. (The demonstration of this principle is left as an exercise at the end of the chapter.) In real populations, there is no real reason to expect that allele frequencies are initially different in males and females, and any observed deviations from HWE are unlikely to be caused by this very transient effect.

   HWE can also be generalized to loci with more than two alleles. Imagine a k-allelic locus with allele frequencies $f_1, f_2, \ldots, f_k$, assumed to be equal among males and females. After one generation of random mating, the genotype frequencies can be obtained by multiplying the appropriate allele frequencies together. So the expected genotype frequency of homozygous individuals with genotype ii is f_i^2 for any allele i, and the genotype frequencies of heterozygous individuals with genotype ij is $2f_i f_j$, for any pair of (different) alleles i and j.

Deviations from HWE 1: Assortative Mating

There are many factors that can cause deviations from HWE equilibrium. First, mating may not be random with respect to genotype. For example, individuals may be more likely to mate with other individuals of the same, or similar, genotype. This is called **assortative mating**. Clearly, if *AA* individuals prefer to mate with other *AA* individuals, *aa* individuals prefer to mate with other *aa* individuals, and *AA* and *aa* individuals rarely mate, there will be fewer heterozygous individuals in the next generation than predicted by HWE. For example, consider a population initially in HWE with an allele frequency of $f_A = 0.5$ and genotype frequencies $f_{AA} = 0.25$, $f_{Aa} = 0.5$, and $f_{aa} = 0.25$. If the population then undergoes one generation of strong assortative mating in which individuals only mate with other individuals of the same genotype, the genotype frequency of the AA genotype will become $f_{AA} = 0.25 + 0.25 \times 0.5 = 0.375$. All offspring of $AA \times AA$ matings (25% of all matings) will be of type AA and a quarter of all offspring of $Aa \times Aa$ matings (50% of all matings) will be of type AA. Using similar arguments we can also find the frequency of aa offspring to be $f_{Aa} = 0.375$, and the frequency of heterozygous offspring will then be $f_{Aa} = 1 - f_{AA} - f_{aa} = 0.25$. The allele frequency is still $f_A = 0.5$ in this example, but there are now only half as many heterozygous individuals as under HWE. If this processes continues for many generations, the population will eventually become entirely depleted of heterozygous individuals.

The opposite situation, where individuals prefer not to mate with individuals of their own genotype, is called *negative assortative* mating or **dis-assortative mating**. Dis-assortative mating can result in numbers of heterozygous individuals in excess of those expected under HWE.

Deviations from HWE 2: Inbreeding

Another mating pattern that can cause deviations from HWE is **inbreeding**. Inbreeding occurs as a result of matings between individuals that are related because they have one or more ancestors in common. The effect of such matings is very much the same as for assortative mating. If these matings are more common than expected under random mating, the proportion of heterozygous individuals will be smaller than under HWE. An extreme type of inbreeding occurs when organisms reproduce by self-fertilization, as many plants do. This type of inbreeding will quickly cause strong deviations from HWE. Assortative mating and inbreeding have similar effects on genotype frequencies: they both increase the proportion of homozygous individuals. The difference is that inbreeding affects the whole genome, while assortative mating affects only those loci that determine the trait or traits that affect mating preference. Assortative mating does not affect genotype frequencies at other loci.

In the early population genetic literature, deviations from HWE were often thought to be a consequence of inbreeding in one way or another. For this reason, we measure deviations from HWE in terms of an inbreeding coefficient (F). We will discuss the inbreeding coefficient in more detail a little later in this chapter.

Deviations from HWE 3: Population Structure

When deriving the HWE theory, we assumed that parents were sampled at random from a population. But what if the population were structured so that it really contained two or more subpopulations? Imagine, for example, a species of lizards inhabiting different islands in the Caribbean.

If we obtained a sample from multiple islands, ignoring this structure of the population, it clearly could not be true that the individuals in the sample had been produced by random mating: individuals from different islands are not likely to mate with each other. Consider the extreme case

Figure 1.2 Two subpopulations with allele frequencies $f_A = 0$ and 1, respectively. In the combined population, obtained by pooling individuals from subpopulation 1 and subpopulation 2, all individuals are homozygous and there is an apparent deficit of heterozygous individuals compared to the HWE expectation.

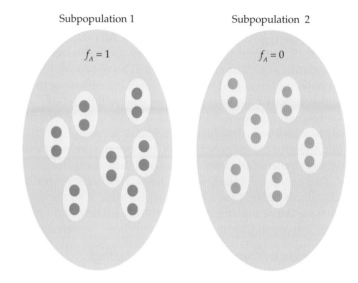

Subpopulation 1 $\quad\quad$ Subpopulation 2

$f_A = 1$ $\quad\quad\quad\quad\quad$ $f_A = 0$

where there are two subpopulations, subpopulation 1 and 2, and the frequency of allele *A* in subpopulation 1 is 100%, while in subpopulation 2, it is 0% (**Figure 1.2**). Even if there is random mating within subpopulation 1 and within subpopulation 2, all individuals will be of either genotype *AA* (subpopulation 1) or *aa* (subpopulation 2). The combined population will very much be out of HWE because it contains only homozygous individuals. Clearly, if there are more than one subpopulation within a larger population (**population structure**), there may be deviations from HWE. This is also true in less extreme cases where allele frequencies differ only marginally between subpopulations. Deviations from HWE will also arise when there are no discrete subpopulations but a continuous spatial distribution of individuals, or in cases when only one subpopulation has been sampled but this subpopulation occasionally receives migrants from another subpopulation. The effect is quite general and is not specific to any particular model of population structure. In real populations, population structure and inbreeding are likely the most important reasons for observations of deviations from HWE. Even relatively small differences in allele frequencies in different subpopulations can cause deviations from HWE. The effects of population structure on deviations from HWE will be discussed in more detail in Chapter 4.

Deviations from HWE 4: Selection

Natural selection occurs when there is differential survival or reproduction among individuals due to their genotypes. It is of such importance in population genetics that we devote three chapters to it. For now, suffice it to say that natural selection also can cause deviations from HWE. Take, for example, the genotype frequencies in the HEXA gene among adults. As individuals homozygous for this disease-causing mutation die before they reach adulthood, the adult population must be slightly out of HWE with a modest excess of heterozygotes. At most, 0.04% of the population is affected by disease, so you would need to examine many thousands of individuals to actually detect this deviation from HWE. Most of the time, we do expect natural selection to be strong enough in humans to cause very severe deviations from HWE. Also worth noting is that deviations from HWE due to selection only can be detected if the population is sampled after selection has been acting. In the case of Tay–Sachs, we do not expect natural selection to cause deviations from HWE among infants.

Some geneticists also include effects of small population sizes and mutations among forces that can cause deviations from HWE. However, as the effect of these factors are extremely small and cause only small random deviations from HWE that do not accumulate over time, we do list them among forces that can cause deviations from HWE.

The Inbreeding Coefficient

Although factors other than inbreeding (such as selection) can cause deviations from HWE, the most common statistic we use to measure deviations from HWE is called the *inbreeding coefficient* (F). To further confuse students, population geneticists have a bad habit of using F to describe the degree to which heterozygosity is reduced both in individuals and in populations as a result of inbreeding. In this book we will use F solely to denote the decrease in heterozygosity in a population beyond that expected under HWE. For a di-allelic locus, we define F as:

$$F = \frac{\left(2 f_A f_a - f_{Aa}\right)}{2 f_A f_a} \tag{1.8}$$

Notice that the first term in the numerator, $2 f_A f_a$, is the proportion of individuals expected to be heterozygous under HWE. So F measures the difference between the expected and the observed heterozygosity, standardized by the expected heterozygosity. If $F = 0$, the population is in HWE, and if $F = 1$, there are no heterozygotes in the population. Also notice that if there are more heterozygotes than expected under HWE, F is negative.

By rearranging Equation 1.4, we find:

$$f_{Aa} = 2 f_A f_a (1 - F) \tag{1.9}$$

which shows that, with this definition, the proportion of heterozygotes in the population is reduced by a factor F from that expected under HWE. If we know the value of F, and the allele frequencies, we can predict the proportion of heterozygote individuals in the population without assuming HWE.

Many plant species are predominantly self-fertilizing and in those species, genotype frequencies are typically far from HWE. For example, in a

Figure 1.3 The flower of wild oats (*Avena fatua*) has both male and female reproductive organs (stamens and pistils) and is capable of self-fertilization, which leads to high levels of inbreeding. Many plants are capable of self-fertilization, but many are not, because they are dioecious (having male and female flowers on separate plants) or because they have evolved other mechanisms to avoid self-fertilization— for example, by separating the flowering times of male and female flowers on the same plant or by evolving genetic self-incompatibility.

population of wild oats, *Avena fatua* (**Figure 1.3**), the genotype frequencies at one locus were found by Marshall and Allard to be $f_{AA} = 0.58$, $f_{Aa} = 0.07$, and $f_{aa} = 0.35$, which obviously deviates from HWE. This species is self-fertile and extensive self-fertilization accounts for the lower frequency of heterozygotes. We can calculate F for this species using the formulas given above. We first find the allele frequencies as $f_A = 0.58 + 0.07/2 = 0.615$, $f_a = 1 - 0.615 = 0.385$. We then find $F = (2 \times 0.385 \times 0.615 - 0.07)/(2 \times 0.385 \times 0.615) = 0.852$.

Testing for Deviations from HWE

If we take a sample from a population, we may randomly tend to get a few more homozygotes or heterozygotes than expected under HWE, even though the population actually is in HWE. To determine if the population is out of HWE, we need a formal statistical test. In such a test, we wish to test the null hypothesis that genotype frequencies follow those predicted by HWE (e.g., Table 1.1 in the di-allelic case). One way of doing this is to use a chi-square test (**Box 1.3**). To perform a chi-square test, we need to obtain the observed and expected values, and to find the degrees of freedom. The genotype counts in the data are the observed values. The expected values are given by the HWE theory and can be calculated by the allele frequencies. There is just one degree of freedom, because there are three categories and two constraints. The first constraint is the same as in the coin toss example in Box 1.1: the genotype counts must add to the total number of observations. The second constraint comes from the fact that the allele frequencies under the expected genotypes should equal the observed allele frequencies.

As an example, consider a locus with the following genotypic counts for forty individuals: $N_{AA} = 20$, $N_{Aa} = 10$, $N_{aa} = 10$. The genotype frequencies are $f_{AA} = \frac{1}{2}$, $f_{Aa} = \frac{1}{4}$, and $f_{aa} = \frac{1}{4}$ and the allele frequencies are then $f_A = \frac{1}{2} + (\frac{1}{4})/2 = \frac{5}{8}$ and $f_a = \frac{1}{4} + (\frac{1}{4})/2 = \frac{3}{8}$. We next need to find the expected

BOX 1.3   The Chi-Square Test

A chi-square test, in the definition used in this book, is used to test the goodness-of-fit of a model using **categorical data**—data that can be presented as the counts of different types of observations, such as the number of different alleles or the number of different genotypes. It also assumes we have a null-hypothesis model that predicts the expected frequencies of each count. It is this model we wish to test. If the observed counts are so different from the expected counts that they cannot be attributed to chance, then the null hypothesis can be rejected (we no longer believe that model to be true). Assume there are k categories of observations, and let the observed counts be $O_1, O_2, \ldots O_k$, and the expected counts under the model be $E_1, E_2, \ldots E_k$. The chi-square test statistic is then calculated as

$$\chi^2 = \sum_{i=1}^{k} \frac{(E_i - O_i)^2}{E_i}$$

If χ^2 is very large, it means that we can reject the null model because the observed and expected counts are more different from each other than expected by chance. But how do we figure out if χ^2 is sufficiently large to reject the null model? It turns out that standard statistical theory shows that, for large amounts of data (under suitable assumptions), χ^2 follows a chi-square distribution with degrees of freedom equal to $k - p$, where p is the reduction in the degree of freedom due to constraints imposed by the model when calculating the expected values. A chi-square test is performed by calculating χ^2, calculating p, and then comparing the value of χ^2 to a chi-square distribution with $k - p$ degrees of freedom. Chi-square distributions with different degrees of freedom are given in Appendix D.

   As an example, imagine that we are interested in testing the null hypothesis that a coin is fair, i.e., that it produces H and T each with probability 0.5 (see Box 1.1). To test this, we toss a coin 50 times and get 29 H and 21 T. Does this show that the coin is biased (not fair)? The expected numbers under the null model of a fair coin are clearly $E_1 = 25$ and $E_2 = 25$, so we get

$$\chi^2 = \frac{(25-29)^2}{25} + \frac{(25-21)^2}{25} = 1.28$$

In this case, the number of categories is $k = 2$, and the only constraint we have on the counts of H and T is that they should sum to 50, implying that $p = 1$, so there is one degree of freedom. Consulting the table in Appendix A we find that the probability of observing a value of $\chi^2 = 1.28$ or larger is close to 0.25. To reject the null model, this probability would need to be much smaller, say less than 0.05, or less than 0.01, so in this case we cannot reject the null hypothesis that the coin is fair. The cut-off value we choose for the probability is called the **significance level**. The choice of significance level is somewhat arbitrary, but most studies choose 0.05 or 0.01.

   Examples of chi-square tests are given throughout this book; the first is in the section on testing HWE.

genotype counts under HWE, given the allele frequencies: $E_{AA} = 40 \times (5/8)^2$ = 15.625; $E_{Aa} = 40 \times 2 \times 3/8 \times 5/8 = 18.75$; and $E_{aa} = 40 \times (3/8)^2 = 5.625$. We then calculate the chi-square statistic (as in Box 1.3) as

$$\chi^2 = \frac{(15.625-20)^2}{15.625} + \frac{(18.75-10)^2}{18.75} + \frac{(5.625-10)^2}{5.625} = 8.711 \quad (1.10)$$

Comparing our observed value of 8.711 to the critical values for a chi-square distribution with one degree of freedom in Appendix 4, we see that the probability of observing a value this high or higher is between 0.01 and 0.001. Using a traditional significance level of 0.05 (critical value = 3.841), we find $p < 0.05$ and reject the null hypothesis of HWE. The genotype frequencies are statistically significantly different from those expected under HWE.

The chi-square test can also be extended to k-allelic loci. The hardest part is to calculate the degrees of freedom. For k alleles there are $k(k + 1)/2$ possible genotypes, i.e., categories in a chi-square test. But there are k constraints, because the allele frequencies in the expected categories have to match the observed allele frequencies. So the degrees of freedom are calculated as $k(k + 1)/2 - k = k(k - 1)/2$.

d.f. for K-allelic loci

Using Allele Frequencies to Identify Individuals

The DNA from an individual can be used to identify the individual. This principle has been used extensively in many connections, most importantly in forensics where DNA is used to determine paternity and to identify someone who was at a crime scene. In the context of forensics, the use of DNA to identify individuals is called **DNA fingerprinting** or **DNA profiling**. In the United States, thirteen microsatellite loci are usually used in forensics. An individual matches a DNA profile if the genotype is identical to the profile at all thirteen loci. But with only thirteen loci, there is some chance than an individual will match a profile by chance alone. To assess the probability (Box 1.1) of a random match, forensic scientists compare the profile to a database of allele frequencies. If the individual carries two alleles for a locus, say allele 1 and allele 2, then the **match probability** is simply $2f_1 f_2$ for a heterozygous individual, and f_1^2 or f_2^2 for a homozygous individual, assuming HW equilibrium. The probabilities calculated for all loci are then multiplied together to provide one final match probability.

There are several problems that arise in the interpretation of match probabilities based on databases. First, the database may not be representative for the population to which the individual belongs. For example, a database of Caucasian individuals may not be appropriate as a reference for an individual from a non-Caucasian background. For this reason, the United States and many other countries have devoted significant efforts to developing large representative databases. Second, the individual may have siblings or other close relatives who also have a high probability of matching the profile. Third, assumptions regarding HW equilibrium and simple multiplication of probabilities among loci may not always be valid. Considerable statistical research has been devoted to these concerns.

References

*Chen J., 2010. The Hardy–Weinberg principle and its applications in modern population genetics. *Frontiers in Biology* 5: 348–353.

*Evett I. W. and Weir B. S., 1998. *Interpreting DNA Evidence: Statistical Genetics for Forensic Scientists.* Sinauer, Sunderland, MA.

Marshall D. R. and Allard R. W., 1970. Maintenance of isozyme polymorphism in natural populations of *Avena barbata. Genetics* 66: 393–399.

*Valverde P., Healy E., Jackson I., et al., 1995. Variants of the melanocyte-stimulating hormone receptor gene are associated with red hair and fair skin in humans. *Nature Genetics* 11: 328–30.

*Recommended reading

EXERCISES

1.1 A researcher examines a locus in which there is a particular C/T polymorphism. She obtains the following genotypic counts: CC: 42, CT: 16, TT: 32. Calculate the genotype frequencies and the allele frequencies in the sample.

1.2 For the data from Exercise 1.1, find the expected homozygosity and the expected heterozygosity, given the observed allele frequencies, and calculate the inbreeding coefficient (F).

1.3 For the data in Exercise 1.1, test if the population is in HWE using a chi-square test at the 5% significance level.

1.4 The proportion of a population suffering from a specific rare genetic disease is 0.02%. Assume that the disease is caused by a single recessive allele and assume that the population is in HWE. How many individuals carry the disease allele in the heterozygous state?

1.5 In another locus there are three alleles—A, C, T—and the genotypic counts in the sample are AA: 10, AC: 10, AT: 5, CC: 20, CT: 5, and TT: 20. Calculate the genotype frequencies and the allele frequencies in the sample.

1.6 For the data from Exercise 1.5, find the expected homozygosity and the expected heterozygosity, given the observed allele frequencies.

1.7 For the data in Exercise 1.5, test if the population is in HWE, using a chi-square test at the 5% significance level.

1.8 An individual has genotype CT for the locus discussed in Exercise 1.1, and genotype AC in the locus discussed in Exercise 1.5. At a crime scene, forensic evidence is found with the exact same (TT, CC) genotype. What is the chance of such a match by random, assuming HWE and the allele frequencies calculated in Exercises 1.1 and 1.5? What is the match probability if the calculation is done using observed genotype frequencies instead?

1.9 Show mathematically that it takes two generations to achieve HWE when the allele frequencies differ between males and females (assume a di-allelic locus).

2 *Genetic Drift and Mutation*

AS PREVIOUSLY MENTIONED, much population genetic theory concentrates on describing the changes of allele frequencies through time. If we understand how and why the frequencies of different alleles change, we have learned a great deal about evolution.

The two most important factors that cause allele frequencies to change through time are natural selection and **genetic drift**. Genetic drift is the random change of allele frequencies in populations of finite size. For example, imagine that the individuals in the small population in Figure 1.1 are randomly mating to produce a new population. Perhaps some individuals leave many offspring, while other individuals leave fewer offspring, not because of natural selection, but because of extrinsic factors not related to genetics. Some individuals might die before they reach the age of reproduction because of events unrelated to their genetic makeup. In addition, some heterozygous individuals will randomly transmit allele A to their offspring, while other will transmit allele a. In a small population, the average number of a's and A's transmitted to the offspring generation by heterozygous individuals may not be exactly equal to the expected number, because of the randomness of the process of Mendelian segregation. As a result of these factors, it is unlikely that the next generation will contain exactly 7 A alleles and 13 a alleles, as the previous generation did. With reasonably high probability, the allele frequencies will have changed between generations. If this process continues over many generations, it can produce large changes in allele frequencies. For example, in a classical experiment demonstrating the effect of genetic drift, Buri (1956) established 107 cages with *Drosophila melanogaster* populations. He propagated the populations by randomly choosing 8 males and 8 females in each generation to mate, and kept track of the frequency of a specific Mendelian allele that could be determined from the eye color of the flies. The initial frequency of the allele was 0.5 in each cage, but after 19 generations, the majority of population cages contained only one allele or the other, i.e., the allele frequency was either 0 or 1. Genetic drift had been acting on the populations to change the allele frequencies.

The Wright–Fisher Model

Population geneticists have developed a number of different models to describe genetic drift. The most common model is the **Wright–Fisher model**, named after the founders of population genetic theory: Sewall Wright and R. A. Fisher. None of the properties of genetic drift we will discuss are particular to the Wright–Fisher model; in fact, most results discussed in the chapter can be derived using very general population genetic ideas and without reference to any particular detailed model. However, it might be easier to understand the principles discussed in the context of a concrete model.

The Wright–Fisher model assumes a haploid population (i.e., a population in which each individual only caries one copy of the genetic material) without sexes, in which each individual reproduces without the need to mate with another individual. Such a model might be appropriate for, say, many bacterial populations. However, it turns out that most of the dynamics of a diploid population with two sexes are almost identical to the dynamics of this haploid model. Because the haploid model is simpler mathematically, we often use it to approximate the diploid model.

The Wright–Fisher model assumes discrete generations, i.e., that an entire population is replaced by its offspring in a single generation. We usually assume that the population is of a constant size $2N$ (to mimic a diploid population of N individuals). Gene copies are transmitted from generation t to generation $t + 1$, by random sampling (independently and with equal probability) of the gene copies in generation t (**Figure 2.1**).

Imagine that you have two bags. One contains $2N$ marbles of different colors, representing different alleles in a population in generation t. The other bag represents generation $t + 1$ and is initially empty. You draw a marble from the first bag, note its color, and add a marble (from an independent stash) of the same color to the second bag—and put the original marble back in the first bag and shake it up. You keep doing this until there are $2N$ marbles in the second bag. The two bags then model the distribution of gene copies in generations t and $t + 1$. The random change in the number of balls of each color (alleles) represents genetic drift. To denote the allele frequencies in different generations, we now express them as functions of time (in generations). In the example of Figure 2.1, the bags contain alleles of two different colors representing the A and the a allele, and the allele frequency changes from $f_A(t) = 7/18$ to $f_A(t+1) = 4/18$.

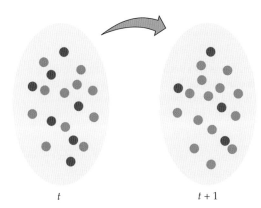

Figure 2.1 An illustration of two generations of a Wright–Fisher population with $2N = 18$ gene copies. In generation t the allele frequency of allele A (red) is $7/18$, but due to genetic drift, the allele frequency is $4/18$ in generation $t + 1$.

The distribution of offspring in generation $t + 1$ is given by what is known as a binomial distribution. This distribution is discussed in more detail in Appendix A.

Genetic Drift and Expected Allele Frequencies

Using the Wright–Fisher model, we can characterize the change in allele frequency mathematically. For example, what is the probability that any particular gene copy in generation $t + 1$ is of type A? Since we are assuming random sampling of the genes in generation t, the chance that any one gene copy we sample is of type A in generation $t + 1$ is simply the frequency of allele A in generation t, $f_A(t)$. We can use this insight to find the expected number of A alleles in the next generation. There are $2N$ gene copies in generation $t + 1$, each is of type A with probability $f_A(t)$, so we expect a total of $2Nf_A(t)$ A alleles. If there are $2Nf_A(t)$ A alleles, then the frequency of the A allele is $2Nf_A(t)/2N = f_A(t)$. Using the mathematical notation from **Box 2.1**, we can write this as

$$E[f_A(t + 1)] = 2Nf_A(t)/2N = f_A(t) \qquad (2.1)$$

That is, the expected allele frequency in generation $t + 1$ is equal to the allele *on average* frequency in generation t. It is important to realize that this is an argument about averages. If we repeat the sampling scheme used in the Wright–Fisher model (infinitely) many times, starting with an allele frequency of $f_A(t)$ each time, then the average of $f_A(t + 1)$ over all the replicates is $f_A(t)$. However, in each individual replicate, it is highly likely that the allele frequency will have changed. Genetic drift leads to changes in allele frequency, but the change is equally likely to favor the A and the a allele. Genetic drift is in this sense blind to the allelic state. As we will later see, this distinguishes it from natural selection, which favors one allele over the other.

BOX 2.1 Expectation

The **expectation** of a random variable is its average value. Imagine rolling a die, and representing the outcome of the roll by the random variable Y. The possible outcomes are 1, 2, 3, 4, 5, and 6; if nobody has tampered with the die, all of these six outcomes are equally likely, i.e.,

$$\Pr(Y = j) = \tfrac{1}{6}, \text{ for } j = 1, 2, \ldots, 6$$

The expectation of Y, $E(Y)$, is then found by taking the average over all possible outcomes and weighting each outcome by its probability; thus,

$$E(Y) = \sum_{j=1}^{6} j\Pr(Y = j) = 1\tfrac{1}{6} + 2\tfrac{1}{6} + 3\tfrac{1}{6} + 4\tfrac{1}{6} + 5\tfrac{1}{6} + 6\tfrac{1}{6} = \tfrac{21}{6} = 3.5$$

The average value of a roll of a die is 3.5. In general, the expected value of a random variable (X) can be found as

$$E(X) = \sum x \Pr(X = x)$$

where the sum is over all possible values of x (the sample space).

There are a couple of mathematical rules for expectations that are worth knowing. If c is a constant (it is not random) then

$$E(cX) = cE(X) \qquad \text{and} \qquad E(c + X) = c + E(X)$$

for any random variable X. For example, if we add 10 to the result of each throw of a die, then the expected value is $10 + 3.5 = 13.5$. Similarly, if we multiply the result by 10 every time we throw the die, the expected value is 35. It is also true that for two random variables, X and Y,

$$E(Y + X) = E(Y) + E(X)$$

You can read more about the concept of expectation in Appendix A.

Patterns of Genetic Drift in the Wright–Fisher Model

What happens when you repeat the Wright–Fisher sampling scheme over many generations? A computer simulation of one hundred generations for ten different populations is illustrated in **Figure 2.2**. Each population evolves according to a Wright–Fisher model, i.e., gene copies are randomly sampled from one generation into the next. In each generation, the allele frequency might change a little. The small changes add up, and after many generations, the allele frequency may have changed significantly. This illustrates one of the most important principles if evolutionary theory: many small changes may result in large evolutionary changes over sufficiently long periods of time.

Also notice that at the end of the simulations, the allele frequency has increased in some populations and decreased in other populations. As

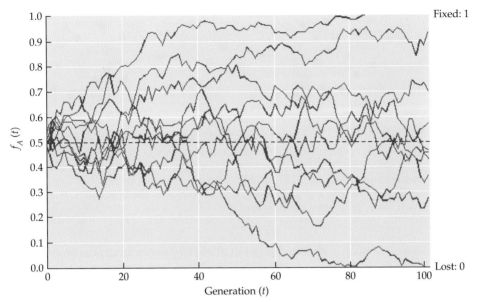

Figure 2.2 The Wright–Fisher model simulated for 10 populations, with $2N$ = 100, over 100 generations (solid lines) for an initial allele frequency of 50%. Allele frequencies change randomly due to genetic drift. The expected (mean) allele frequency is shown by the dashed line.

the change in allele frequency is random and is equally likely to favor the A and the a allele, we would expect this to happen. In some cases, the A allele frequency reaches 1 or 0. When this happens, we say the allele has become fixed [$f_A(t) = 1$] or lost [$f_A(t) = 0$]. Because we assume no mutation, when an allele first has become fixed or lost, its frequency cannot change anymore. When an allele has been lost from the population, its frequency will remain at 0% in perpetuity. Likewise, if the frequency of allele A is 100% in generation t, it must be in generation $t + 1$.

In the absence of recurrent mutation, an allele must eventually become fixed or lost: it cannot be maintained in the population forever. This can be shown mathematically by noticing that in every generation there is some positive (bounded) probability that the allele will become lost or fixed. Intuitively, it should make good sense that as genetic drift continues for many generations, eventually the allele frequency will reach either zero or one.

Effect of Population Size in the Wright–Fisher Model

How fast can genetic drift change allele frequencies? The answer to this question depends on the population size, N. Consider a very small population with, say, $N = 10$. In such a population, there is a reasonable chance that the allele frequency will change from, say, $5/10$ to $3/10$ or from $5/10$ to $7/10$ in one generation. In fact, a small calculation can be done to show that the

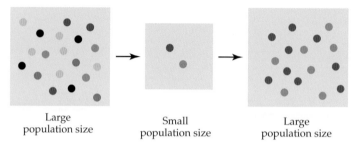

Large
population size · Small
population size · Large
population size

Figure 2.3 An illustration of the effect of a bottleneck in population size on genetic variability. The initial population has a high degree of variability, illustrated by the variety of colors of the balls in the box. The population goes through a bottleneck—a temporary decrease in size—and after the bottleneck, there are many fewer different alleles present.

chance of such a strong change in allele frequency is approximately 0.34 when the initial allele frequency is 5/10. Now consider a population of 1000 individuals, and a similar change in allele frequency, i.e., a change in allele frequency from 0.5 to 0.3 or 0.7 in one generation. The chance of such a strong change is less than 2×10^{-37}. Large changes in allele frequency are unlikely in large populations, but happen more easily by chance in small populations. In other words, genetic drift works much faster in small populations than in large populations.

The effect of population size on genetic drift has important implications for our understanding of natural populations. Many populations experience bottlenecks in the population size—short periods of time when the population size is very small and many alleles are either fixed or lost in the population (**Figure 2.3**). As a consequence, much of the genetic variation in the population is lost. For example, a century ago the northern elephant seal (*Mirounga angustirostris*) (**Figure 2.4**) went through a period in which it had a population size of only about 2–20 individuals. They were hunted nearly to extinction for their blubber. Today, the population has rebounded to more than 175,000 individuals. However, due to the historical bottleneck in population size, the northern elephant seal has drastically reduced variation. In a study of more than 100 individuals, Weber et al. (2000) found only two different mtDNA haplotypes (different types of DNA sequences). When they sequenced DNA from museum specimen collected before the bottleneck, they found many more haplotypes in the sequenced individuals. The haplotype heterozygosity, a common measure of genetic diversity was reduced from 0.9 before the bottleneck to 0.41 after the bottleneck.

A bottleneck in population size may happen when a new population, or species, is formed as a few individuals become isolated from the rest of the population. The reduction in variability caused by a bottleneck in population size during the founding of a new population is called the **founder effect**. Founder effects place a special role in theories of specia-

Figure 2.4 Elephant seals experienced a drastic reduction in population size around the turn of the 19th century. Today, the population size has recovered, but they still suffer from reduced genetic variability.

tion processes; some of these theories posit that genetic divergence after speciation may be helped along by the strong effect of genetic drift in the founders of a population.

Mutation

According to the description of genetic drift given by the Wright–Fisher model, one might expect natural populations to contain no genetic variation, because all alleles eventually become fixed or lost, but that is obviously not the case. In real populations, new mutations arise to produce new genetic variation that genetic drift can act on. Mutations come in many forms. From basic genetics, you might be familiar with mutations such as deletions (in which part of a DNA sequence is removed), insertions (in which new DNA is inserted into a chromosome), inversions (in which the orientation of a piece of DNA is inverted), translocations (in which a piece of DNA is moved from one chromosome to another), and point mutations (in which one nucleotide is replaced by another). We encourage students who are not familiar with the molecular basis of mutation to consult a textbook on basic genetics or molecular biology to review these topics.

Any of the mutations mentioned above can be modeled with a di-allelic model, in which one allele represents the presence of the mutation, and one allele represents the absence of the mutation. The population genetics of an inversion mutation is the same as that of a point mutation, so we can investigate the population genetic theory without reference to the molecular identity of the mutation. However, if multiple mutations can occur in the same location, the details of the models of mutation are important. A simple di-allelic model may not be sufficient to account for the transmission of the mutation, and we may need more complicated models. But for now, we will use a simple model with two alleles, A and a.

Effects of Mutation on Allele Frequency

We have already seen that genetic drift might change allele frequencies. But what is the effect of mutation on allele frequencies? Consider again the Wright–Fisher model, but assume that the a allele in each individual randomly mutates to A with probability μ in each generation. The parameter μ is also called the **mutation rate**. What is the expected allele frequency in generation $t + 1$ now if the allele frequency in generation t is $f_A(t)$? The probability that the parent of any individual in generation $t + 1$ was of type A is $f_A(t)$. However, even if the parent carried allele a, there is a probability μ that the offspring is of type A. The expected allele frequency in the next generation is, therefore,

$$E[f_A(t + 1)] = f_A(t) + \mu f_a(t) \tag{2.2}$$

The allele frequency of allele A is expected to increase by a fraction $\mu f_a(t)$ from generation t to generation $t + 1$. If this process continues for a long time, and there are no other forces affecting allele frequencies, eventually all individuals in the population will be of type A.

If mutation occurs in both directions, i.e., mutations occur at rate $\mu_{a \to A}$ from a to A and at rate $\mu_{A \to a}$ from A to a, the expected frequency in the next generation is

$$E[f_A(t + 1)] = (1 - \mu_{A \to a})f_A(t) + \mu_{a \to A}f_a(t) \tag{2.3}$$

In the absence of other forces such as genetic drift and selection, an equilibrium will eventually be established. The equilibrium value is attained when there is no change in allele frequencies: when $E[f_A(t + 1)] = f_A(t) = f_A$, that is, when

$$(1 - \mu_{A \to a})f_A + \mu_{a \to A}f_a = f_A \tag{2.4}$$

Substituting $(1 - f_A)$ for f_a, and rearranging this equation a bit, we find that at equilibrium,

$$f_A = \frac{\mu_{a \to A}}{\mu_{a \to A} + \mu_{A \to a}} \tag{2.5}$$

For example, if $\mu_{A \to a} = \mu_{a \to A}$ then the equilibrium value is $f_A = \frac{1}{2}$, as you might expect.

Mutation rates are highly variable among different organisms; some bacteria have very high mutations rates, while most eukaryotes have quite small mutation rates. Typically, the mutation rate for point mutations in

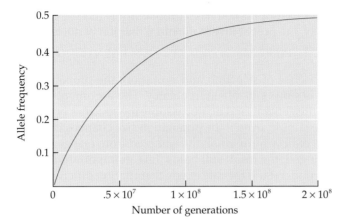

Figure 2.5 The allele frequency of an allele with initial frequency 0 and with a mutation rate toward the allele and away from the allele of $\mu_{A \to a} = \mu_{a \to A} = 10^{-8}$, a typical value for vertebrates. Notice that if mutations were the only force affecting allele frequencies, it would take 10^8 generations for the population to reach equilibrium.

higher organisms is on the order of 10^{-7} to 10^{-9}. So mutation is a very weak force in higher organisms. If there were no genetic drift, we would have to wait a very long time for mutations alone to change allele frequencies, and an even longer time for equilibrium to be reached. This is illustrated in **Figure 2.5**, which plots Equation 2.4 for values of the mutation rate associated with vertebrates.

Most of the time, we can safely ignore recurrent mutations occurring at the same site (position in the DNA sequence) when we model changes of allele frequencies. Other forces, such as natural selection and genetic drift, will be more important.

Probability of Fixation

As previously noted, in the absence of mutation, any allele must eventually be lost or fixed. Similarly, in a haploid population, either all or no individuals in the population will eventually be descendants of any particular individual from generation t; eventually, only one gene copy from generation t will be present in the descendant population. As there is no selection that can increase or decrease the chance that any particular gene copy will be the lucky one, the probability must be the same for all gene copies. And as the probabilities for all individuals must sum to one, the probability for any one individual must be $1/(2N)$. So the probability that an allele of frequency $1/(2N)$ goes to fixation is simply $1/(2N)$.

We can now answer the more general question of the probability of fixation of an allele of frequency $f_A(t) = N_A/(2N)$. Each of the N_A copies of the *A* allele in the population has a probability $1/(2N)$ of going to fixation. Consequently, the probability of eventual fixation of allele A is

$$\Pr(\textit{fixation of allele A}) = N_A \times 1/(2N) = f_A(t) \qquad (2.6)$$

In the absence of selection and mutation, the probability of fixation of an allele is simply its allele frequency. The arguments leading to this result are quite general, and do not assume the specifics of a Wright–Fisher model. For example, even if the population size changes through time or if the specifics of how individuals are sampled from one generation into the next are changed, the result still holds. It is an important result, because it will help us understand the rate at which mutation differences accumulate between species.

Species Divergence and the Rate of Substitution

Armed with the knowledge that the probability of fixation equals the allele frequency, we can now easily derive the rate at which mutations accumulate between species. We call this the **rate of substitution**. We use the word *substitution* to indicate mutations that have gone to fixation. In this context, if we equate a population to a species, we can think of a substitution as a fixed mutation between species.

Assume that the mutation rate is μ—that is, in each generation μ new mutations occur in each gene copy. The mutation rate can be reported per site, per gene, per chromosome, per genome, or in other units. In any particular application, these units are important. However, the general results we will derive here do not depend on the units, so we will not explicitly keep track of them. As there are $2N$ gene copies, the number of mutations entering the population each generation is $2N\mu$. Each initially has a frequency of $1/(2N)$, so the probability that any one goes to fixation is $1/(2N)$, and the expected number of mutations each generation that eventually will go to fixation is

$$2N\mu \times 1/(2N) = \mu \qquad (2.7)$$

The rate of substitution is, therefore, simply the mutation rate. Perhaps somewhat surprisingly, the rate of substitution does not depend on the population size.

The Molecular Clock

The result of the previous section tells us that if there is no selection, and if the mutation rate is low enough that it does not affect allele frequencies much, the rate of substitution should be constant in time if the mutation rate is constant in time. Mutational differences between species should accumulate at a constant rate. This means that mutations can be used to date divergence times between species. For example, imagine that we compare a particular gene, or genomic region, in two species and count the number of sites where the nucleotides are different from each other in the two species (d = the number of nucleotide differences). We will assume that each new mutation that occurs in the history of the species creates one new nucleotide difference (this may not always be realistic, as two mutations could hit the same site). If the nucleotide mutation rate in the region is μ per generation, then we would expect the average number of nucleotide differences separating the two sequences to be

$$E[d] = 2\mu t \qquad (2.8)$$

assuming that they have been diverging from each other for t generations (**Figure 2.6**). Notice the factor of 2, which is necessary because mutations accumulate in both species.

This equation is used for estimating divergence times. For example, if we are told that two species, say species A and B, are separated by d_{AB} nucleotide differences, then by rearranging Equation 2.8, we can estimate t_{AB} as $2\mu/d_{AB}$. However, we would first need to know the value of μ. To estimate μ we could examine some other species, say C and D, for which the divergence, t_{CD}, is known, and estimate μ as $d_{CD}/(2t_{CD})$.

The **molecular clock**, originally conceived by E. Zuckerkandl and L. Pauling, has been used in thousands of applications over the past forty years to estimate the divergence times between species. However, its use comes with a warning. Two central assumptions are that the mutation rate is constant and that there is no natural selection. Neither of these assumptions is particularly realistic. Mutation rates seem to change between species, especially very divergent species, and natural selection seems to be important in the evolution of DNA sequences in most organisms. However, between closely related species, the use of the molecular clock is at times surprisingly accurate. Also, much research has been devoted to finding statistical methods to correct for varying mutation rates.

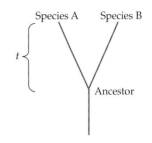

Figure 2.6 Two species, A and B, diverged from a common ancestral species t generations ago.

Dating the Human–Chimpanzee Divergence Time

Based on paleontological evidence, it has been argued that cercopithecoid primates, such as the rhesus macaque (*Macaca mulatta*), and hominoids, such as humans and chimpanzees, separated from each other 25 million years ago. The rhesus macaque genome sequencing consortium found that humans and macaques differ in 7% of the nucleotide positions of the part of their genomes that can be directly compared. The chimpanzee sequencing consortium found that humans and chimpanzees differ in approximately 1.2% of these positions. Using the macaque–human comparison, we can infer the mutation rate to be $(0.07/2)/25 \times 10^6 = 1.4 \times 10^{-9}$ per year. Assuming a molecular clock, we can then date the human–chimpanzee divergence to be $0.012/(2 \times 1.4 \times 10^{-9}) = 4.3$ million years ago. While such a short divergence time is compatible with some recent estimates, it is considerably smaller than other estimates, which suggest a divergence time of 5–6 million years. It also seems to conflict with the fact that the oldest fossil with clear humanlike features (*Ardipithecus ramidus*) is 4.4 million years old.

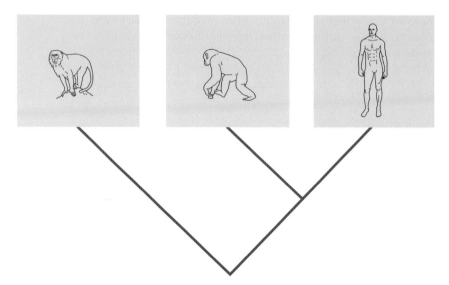

Figure 2.7 The evolutionary tree relating humans, chimpanzees and the rhesus macaque. The human–chimpanzee divergence time is 5–6 million years, and the divergence time between the rhesus macaque and the humans–chimpanzee clade is about 25 million years.

How can we explain the discrepancies? One likely explanation is that the generation time has changed. Humans and chimpanzees both have longer generation times than the rhesus macaque (**Figure 2.7**). Mutations in the germline (heritable mutations) are thought to occur during meiosis, so we might expect that the mutation rate is relatively constant per generation. The mutation rate per year would, therefore, be higher in the macaque lineage, leading to an underestimate of the human–chimpanzee divergence time. Other explanations might include effects of natural selection, changes in the biological mutation rate through time, bioinformatical or experimental errors, and the possibility that the cercopithecoid–hominoid split occurred more than 25 million years ago, in which case we would have overestimated the mutation rate.

References

Buri P., 1956. Gene frequency in small populations of mutant *Drosophila*. *Evolution* 10: 367–402.

*King J. L. and Jukes T. H., 1969. Non-Darwinian evolution: Random fixation of selectively neutral mutations. *Science* 164: 788–798.

*Rhesus Macaque Genome Sequencing and Analysis Consortium, 2007. Evolutionary and biomedical insights from the rhesus macaque genome. *Science* 316: 222–234.

*Weber D. S., Stewart B. S., Garza J. C., Lehman N., 2000. An empirical genetic assessment of the severity of the northern elephant seal population bottleneck. *Current Biology.* 10:1287–1290.

*Recommended reading

EXERCISES

2.1 At a particular locus, there are four nucleotides segregating: A, C, T and G, with frequencies in the population of 0.1, 0.2, 0.2, and 0.5, respectively. Assume that the population evolves in accordance with a Wright–Fisher model, with no new mutations. What is the probability that nucleotide A at this locus will eventually be lost from the population? What is the probability that either A or C will go to fixation?

2.2 An insertion mutation is segregating in a population at a frequency of 10%. Each generation, in each individual carrying the mutation, the probability that this mutation is reverted to the ancestral form by a deletion mutation is 10^{-6}. The mutation also occurs *de novo*, in individuals not already carrying it, at a rate of 5×10^{-6} per generation. If there is no genetic drift or selection, what is the expected frequency of the mutation after one generation?

2.3 What is the expected equilibrium frequency of the mutation discussed in Exercise 2.2?

2.4 The mutation rate in a particular gene is 1×10^{-9} per generation per base pair (bp). The gene is 800 bp long. Assume that both humans and chimpanzees have a generation time of twenty years, and that each mutation will create a new nucleotide difference between chimpanzees and humans. If the divergence time between humans and chimpanzees is 6 million years, how many nucleotide differences in this gene would you expect to observe between humans and chimpanzees?

2.5 The divergence time between two species of fish (species A and B) is 20 million years. In a particular gene, they differ by 29 nucleotide differences. The number of nucleotide differences between species A and another species (species C) is 12. Based on this information, and the assumption of a molecular clock with a constant generation time, provide an estimate of the divergence time between species A and C.

2.6 Consider a standard neutral Wright–Fisher model with population size $2N$. What is the expected number of offspring of a particular individual in the next generation?

2.7   For the model in Exercise 2.6, write a formula for the probability that a particular individual from generation t leaves no descendants in generation $t + 1$.

2.8   For the model in Exercise 2.6, write a mathematical formula for the probability that a mutation of frequency p is lost from the population within one generation.

2.9   For the model in Exercise 2.6, what is the probability that two individuals in generation $t + 1$ both have the same parent in the previous generation?

3 Coalescence Theory: Relating Theory to Data

IN CHAPTER 2, we developed a theory of genetic drift based on the Wright–Fisher model. Ultimately, we would like to be able to relate the theory to data, so we can use real DNA sequence data to learn more about the populations from which data have been sampled. For example, Wall et al. (2008) sequenced DNA from the X-chromosomes of various human populations. They found that two Europeans differed, on average, in 0.08% of the sites (positions in the DNA sequence). However, individuals from African populations differed in 0.12% of DNA sites. What do these numbers tell us about the two populations? We can use the Wright–Fisher theory to answer this type of question although it is often difficult and mathematically awkward to do so. However, in the early 1980s a new mathematical theory of population genetics was developed by mathematicians such R. C. Griffiths and J. F. C. Kingman, and biologists such as R. R. Hudson and F. Tajima. This theory, called **coalescence theory**, is based on models such as the Wright–Fisher model. But instead of modeling changes of allele frequencies forward in time, it considers a sample and the genealogical history of the sample. Coalescence theory is tremendously useful for analyzing real data, because it provides a population genetic theory that relates directly to data sampled from a population. In this chapter, we will explain some of the foundations of coalescence theory, and show how it can be used for making inferences about real populations.

Figure 3.1 A sample of three (haploid) individuals in generation $t + 1$. Two individuals have the same parent in generation t, while the third individual has another parent.

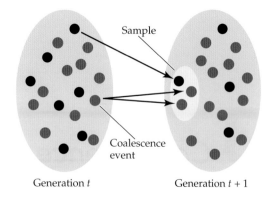

Sample

Coalescence event

Generation t Generation $t + 1$

Coalescence in a Sample of Two Chromosomes ($n = 2$)

In the Wright–Fisher model, each individual gene copy in generation $t + 1$ chooses its parent with equal probability among all parents in the previous generation. Imagine tracking the ancestry of a sample from the population between two generations as in **Figure 3.1**.

   If two individual gene copies in our sample have the same parent in the previous generation, we say that the **ancestral lineages** representing these two individuals have **coalesced**. They have a **common ancestor**, and a **coalescence event** (or **coalescent**) has occurred. Now, in most cases, a pair of individuals in generation $t + 1$ do not have a common ancestor in generation t. However, the two parents of the individuals may have had a common ancestor in generation $t - 1$, or maybe a shared grandparent as a common ancestor in generation $t - 2$. Mathematically, it can easily be shown that for this model, eventually, the two lineages will find a common ancestor. We can imagine the ancestry of two sampled individuals as depicted in **Figure 3.2**.

   The ancestry of an individual gene copy is represented by a line (also called an **edge** in the terminology of graph theory) in the diagram. We can think of the ancestry of the two samples as a tree where each lineage in the tree represent ancestry of a gene copy. The time until the two lineages find a **most recent common ancestor** (**MRCA**), i.e., the time until their ancestors for the first time share the same parent, is of central interest in population genetics. The time to the most recent common ancestor is called the **coalescence time**. To find it, we first have to find the probability that two individuals have the same parent in the immediately previous generation. This probability is simply the probability that the second individual has the exact same parent as the first individual. Imagine throwing a die two times. The chance that the second throw results in the same outcome as the first throw is ⅙, and is independent of the outcome of the first throw. If the result of the first throw is 1, then the chance that the second throw is also

[handwritten notes in margin:] where lineage represents ancestry of a gene copy

Present   Coalescence tree

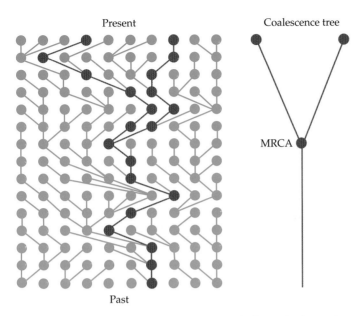

Past

Figure 3.2 The ancestry of a sample of two individuals. Each dot in the array on the left represents an individual in a haploid population. Each horizontal line of dots represents a generation, so that the panel shows the genealogy for the entire population (10 haploid individuals) for 15 generations. Lines in the graph represent descent (parent–offspring relationships). The ancestry of the two sampled individuals is highlighted in red. A most recent common ancestor (MRCA) for the two individuals is found after 6 generations when the two lineages coalesce. The panel to the right shows the resulting coalescence tree for the two individuals. The dots in this graph are called nodes. The two on the top are leaf nodes representing the sampled (haploid) individuals. The one further down in the tree represents the MRCA of the two individuals.

1 is ⅙. If the result of the first throw is 2, then the chance that the second throw is also 2 is ⅙, and so on. Similarly, because there are 2N potential parents "chosen" with equal probability, the probability of two individuals having the same parent in the previous generation is

Pr(2 *gene copies have the same parent in previous generation*) $= 1/(2N)$ (3.1)

The probability that the two gene copies did not have the same parent (did not coalesce) in the previous generation is then $1 - 1/(2N)$. The same will be true for all previous generations in the past. The probability that the two gene copies did not have the same parent in any of the past r generations is then obtained by multiplying together the probabilities for each of the r generations:

Pr(2 *gene copies do not find a common ancestor in r generations*)
$$= [1 - 1/(2N)]^r \quad \# \text{ of generations} \quad (3.2)$$

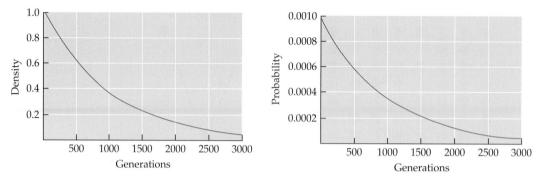

Figure 3.3 The distribution of the number of generations until two lineages find an MRCA (the coalescence time) for a model with discrete generations, the Wright–Fisher model, (right) and for the continuous approximation (left). In both cases a population size of $2N = 1000$ is assumed. Notice the strong similarity between these two distributions of coalescence times. For $2N = 1000$ the continuous distribution clearly provides a very close approximation to the discrete distribution arising under the Wright–Fisher model.

Likewise, the chance of not finding any common ancestor in generation $r - 1$, but then finally finding the first common ancestor in generation r is

$$\text{Pr}(2 \; gene \; copies \; find \; a \; first \; common \; ancestor \; in \; generation \; r)$$
$$= [1 - 1/(2N)]^{r-1} \, [1/(2N)] \quad \quad (3.3)$$

This equation is important because it gives us the probability distribution of the time to the most recent common ancestor (the coalescence time) in a sample of size $n = 2$. The distribution is given by a geometric random variable, shown in the right panel in **Figure 3.3**, and discussed in more detail in Appendix A.

Coalescence in Large Populations

Assumption → Most coalescence theory is based on approximations, assuming that population sizes are large. If we consider the limit of an infinitely large population, i.e., $N \to \infty$, a number of calculations simplify considerably. Considering $N \to \infty$ does not mean that there is no genetic drift, or that we cannot capture much of the evolutionary dynamics of small populations, it just means that we ignore some mathematical details that really matter only if the population is small. These details depend strongly on the specific models. For example, a popular alternative to the Wright–Fisher model is the Moran model, named after the Australian statistician P. Moran. The Moran model does not assume that all individuals are replaced instantaneously in the population, but rather that they are replaced one individual at a time. For small populations, the Moran and the Wright–Fisher models have somewhat different dynamics. But for large population sizes, their coalescence processes are identical (if time is scaled appropriately). In fact, the coalescence process that we will derive in this chapter applies to a much larger set of models than just the Wright–Fisher model.

Using the approximation of infinite population sizes has the effect that time is measured continuously instead of in discrete generations. It also becomes more convenient to measure time in terms of $2N$ generations. We obtain this change in how we measure time by setting $r = 2Nt$, where t measures time in terms of $2N$ generations. We then subsequently consider the limit of $N \rightarrow \infty$. Using a result from calculus (Euler's limit definition of the exponential function), we find that the probability that two gene copies do not find a common ancestor in $2Nt$ generations becomes

$$[1 - 1/(2N)]^{2Nt} \rightarrow e^{-t} \text{ as } N \rightarrow \infty \qquad (3.4)$$

Readers unfamiliar with this calculus fact may at this point just choose to believe us, or they can consult a calculus textbook to learn more. The result means that as N becomes large, the distribution of the coalescence times follows an **exponential distribution** with mean 1 as shown in red in Figure 3.3 (also see Appendix B). We might write $E[t] = 1$, if t is the coalescence time. But as time is now measured in $2N$ generations ($t = r/2N$), the mean (expected) time to coalescence is actually $2N$ generations.

To say that the distribution is exponential with a mean of 1 when scaling time in terms of $2N$ is the same as saying that there is a constant rate of coalescence of 1 per $2N$ each generation. The expected time you have to wait until the coalescence event is 1 divided by the coalescence rate, in this case $1/1 = 1$. These concepts might be a bit difficult to understand for readers not familiar with probability theory, but an analogy might help. Imagine standing on the sidewalk on a street in Manhattan late at night waiting for the first empty taxi cab to come by. Assume that on average, there are two empty taxis coming by every hour. The expected time you have to wait for the first taxi is half an hour (although the actual time might be smaller or larger). Similarly for the coalescence process: we look backward in time and wait for the first coalescence event to happen. The rate of coalescence is 1 per $2N$ generations, and the mean time we have to wait until the first coalescence event is thus also 1 when measuring time in terms of $2N$ generations (or $2N$ when measuring time in terms of generations).

If empty taxis in Manhattan arrive at a constant rate, that means that the chance that one will arrive within the next ten minutes is the same whether we have been waiting two minutes or twenty minutes. The same is true for the coalescence process. That the time to coalescence is exponentially distributed tells us that the chance a coalescent event happens in any particular time interval, given that it has not happened before the time interval, is the same for all time intervals. This does not mean that the chance that the coalescence event happened in an interval, say, between 100 and 110 generations ago is the same as the chance of its happening in a time interval between generation 0 and 10; because if it has already happened within the first 100 generations, it could not also happen after that; there is only one coalescence event. The exponential density (Appendix B) is a decreasing function (see Figure 3.3) implying that more recent coalescence times are more likely, although the rate at which the coalescence event happens (given that it has not already happened) is constant in time.

The random process described here, in which we follow the lineages backward in time until a most recent common ancestor has been found, is called a **coalescence process**. Although the mathematics of the process is rather abstract, and perhaps difficult to understand at first, it should make good intuitive sense that the expected coalescence time is 2N generations. If the chance of rolling 2 on a die is ⅙, you need, on average, to roll the die six times before you roll a 2. Similarly, the chance of a coalescence event in any particular generation is $1/(2N)$, so you must, on average, wait 2N generations for the first coalescence event to happen. It is important to realize that even though the mean coalescence time is 2N, there is considerable variability in the coalescence time (see Appendix B). Because of the inherent randomness in the coalescence process, the time to coalescence may often be much smaller or much larger than the mean.

We now have a convenient description of the genealogical history (coalescence process) of a sample of size $n = 2$. Using this result, it is very easy to derive various sample properties such as the expected heterozygosity, the expected number of segregating sites (variable sites), etc., for a sample of size $n = 2$. This will allow us to connect observations from real data with the population genetic models. The critical reader might now object that we have used a very simple haploid model, but that many organisms of interest (such as humans) are diploid and have two sexes. So is this coalescence process relevant for such species? The answer to this question is yes. The coalescence process in a large randomly mating diploid population with two sexes is the same as that in the simple haploid model. The dynamics of the populations differ when observed over a few generations, but when we consider large populations observed over many generations, this difference tends to vanish. The coalescence model is in this way more general than the Wright–Fisher model.

assumption →

Mutation, Genetic Variability, and Population Size

As before, we will assume that new mutations occur with probability μ in each generation. This means that in r generations we expect μr mutations. If we again measure time by 2N generations—where $t = r/(2N)$—we expect $2N\mu t$ mutations on a lineage of length t. As there are two lineages, the expected number of mutations occurring in the history of a sample of size $n = 2$ is $2 \times 2N\mu t = 4N\mu t$. We also know that the expected coalescence time is $E[t] = 1$, so the expected number of mutations separating two gene copies is simply $4N\mu$. Population geneticists are so excited about this result that they have devoted a Greek letter entirely to this, and commonly write $\theta = 4N\mu$.

These results point to a simple relationship between the amount of genetic variability and population sizes. Small populations have on average short (more recent) coalescence times, and therefore harbor less genetic variability, because fewer mutations have accumulated. In large populations, the average coalescence times are longer (more ancient), and the populations, therefore, harbor more genetic variability (more mutations).

The expected number of mutations occurring in a lineage during any time interval of length τ (measured in $2N$ generations) is simply $2N\mu\tau = \tau\theta/2$. We can, therefore, think of the data generated by a coalescence process producing a coalescence tree, and a subsequent process in which mutations are distributed evenly across the lineages of the tree at rate $\theta/2$, so that the expected number of mutations in any segment of the tree of length τ is $\tau\theta/2$. This decoupling of the coalescence process and the mutation process greatly helps to simplify many calculations using coalescence theory. However, if natural selection is acting, the coalescence process and the mutation process are no longer decoupled. For that reason, much of coalescence theory does not easily extend to models with selection. A few simple cases will be discussed in Chapter 8.

Infinite Sites Model

The final set of assumptions we need in order to relate data to the population genetic models is related to the way that mutations affect patterns of variability. There are a number of different population genetic models of mutation, each appropriate for a different type of data. For DNA sequence data, i.e., data in which a gene, or genomic region, has been sequenced, we often use the **infinite sites model**. The basic assumption of the infinite sites model is that each new mutation creates a new variable site, i.e., that *assumption* each new mutation hits a new site in the sequence, such that no site experiences more than one mutation. This assumption is based on the idea that the sequence is infinitely long (has infinitely many sites), so that the chance that two mutations hit the same site is essentially zero. Consider data such as these:

```
Sequence 1 aggtatgcta gaaccctaga aagacacaga gatagacaag

Sequence 2 aggtatgcta gaaacctaga tagacacaga gatagacaag

Sequence 3 aggtatgcta gaaacctaga tagacacaga gatagacaag

Sequence 4 aggtatgctg gaaccctaga tagacacaga gatagacaag

Sequence 5 aggtatgctg gaaccctaga tagacacaga gatagacaag
```

Imagine that these sequences (each consisting of 40 nucleotide sites) have been obtained from various individuals in a population. In these data, the only sites in which some of the individuals differ are sites number 10, 14 and 21 (bold). Such sites are called **segregating sites**, or **single nucleotide** *SNPs = segregating* **polymorphisms (SNPs)**. Under the infinite sites model, there can be at most *sites* one mutation occurring at each site; therefore we can immediately deduce that only three mutations occurred in the ancestry of these sequences. Furthermore, as the model does not distinguish between the different nucleotides, A, C, T, and G, and does not care about invariable sites, we can simply represent the data as a binary matrix of the variable sites:

Sequence 1   0 0 0

Sequence 2   0 1 1

Sequence 3   0 1 1

Sequence 4   1 0 1

Sequence 5   1 0 1

The labelling of nucleotides with zeros and ones is arbitrary, designating only whether a sequence carries the same or a different allele compared to a chosen reference. The infinite sites model provides a reasonable approximation to a full model of mutation between A, C, T, and G's, if the rate of mutation is so low that the probability of more than one mutation in the same site is very low. DNA sequences with different mutations are different haplotypes. In the example above, there are five DNA sequences, but only three different haplotypes.

The Tajima's Estimator

To estimate θ, we can use the assumption of an infinite sites model and the expected number of mutations separating two individuals. An **estimate** is an educated guess of the true value of a parameter based on information obtained from data. In our case, the parameter is θ and the data are the DNA sequences shown above. The data can be summarized in different ways. A popular way of summarizing DNA sequence data is in terms of the **average number of pairwise differences**, or π. The value of π is obtained by calculating the number of sites in which each pair of sequences differ, and then taking the average among all pairs of sequences. We can write this as

avg # pairwise differences *

$$\pi = \frac{\sum_{i<j} d_{ij}}{n(n-1)/2} \tag{3.5}$$

where n is the number of sequences that have been sampled, d_{ij} is the number of nucleotide differences between sequence i and j, and the sum in the numerator is over all possible distinct pairs of i and j. The denominator is the number of such pairs.

In our example, there are five sequences. For five sequences there are $5 \times 4/2 = 10$ different pairs. Compare the first sequence with the four other sequences, then compare the second sequence with the three remaining sequences, etc. You will realize that there are $4 + 3 + 2 + 1 = 10$ [$= n(n-1)/2$ for $n = 5$] different comparisons. Comparing the first and the second sequence, we find two differences; comparing the first and the third sequence, we find again two differences, and so on. Continuing like this we find that the average number of pairwise differences in this particular case is $\pi = (2 + 2 + 2 + 2 + 0 + 2 + 2 + 2 + 2 + 0)/10 = 1.6$

Under the infinite sites model, each mutation creates a new variable site, so the number of nucleotide difference between two sequences equals the number of mutations occurring in the ancestral lineages of the two sequences before coalescence. The expected number of nucleotide difference between two sequences is simply the expected number of mutations, which we have already shown to be equal to θ. For two randomly sampled sequences, i and j, under the infinite sites model and our standard coalescence model, we have

$$E[d_{ij}] = \theta \qquad (3.6)$$

The number of pairwise differences should thus be a reasonable way to estimate θ. The number of differences between each particular pair of sequences should on average be θ, but we want to use the information from all pairs of sequences, so it makes good sense to use an average over all pairs as an estimate of θ.

The value of π provides an unbiased estimate of θ. That the estimate is unbiased means that on average, it will provide the true value (although any particular value might be off), i.e., $E[\pi] = \theta$. We can show this to be true using the following arguments:

$$E[\pi] = E\left[\frac{\sum\limits_{i<j} d_{ij}}{n(n-1)/2}\right] = \frac{\sum\limits_{i<j} E[d_{ij}]}{n(n-1)/2} = \frac{\theta n(n-1)/2}{n(n-1)/2} = \theta \qquad (3.7)$$

The first equality sign follows from the definition of π, the second follows from the properties of expectation discussed in Box 2.1, the third from the fact there are $n(n-1)/2$ pairwise comparisons and from the previous equation given above, and the last is trivially true.

We now have a convenient statistical method for estimating θ. A quantity for estimating parameters from the data is called an **estimator**. The estimator of θ based on π is sometimes called **Tajima's estimator** after the Japanese population geneticist F. Tajima. We denote an estimator of a quantity such as θ, by putting a circumflex, or "hat" on θ. We could, for example, write $\hat{\theta}_T = \pi$. The subscript T indicates that this is the particular estimator of $\hat{\theta}_T$ named after F. Tajima. It is important to distinguish the parameter θ, with a true value we may never know, from the estimator $\hat{\theta}_T$, which takes on a different value for each data set we consider.

The Concept of Effective Population Size

So far we have assumed a very simple population genetic model—the Wright–Fisher model—and its extension to a coalescence process. This model is a considerable simplification of how real populations evolve. For example, most natural populations do not maintain a constant population size, many do not have discrete generations, and almost all have some degree of population structure. However, for purposes of mathematical simplicity, we tend to want to interpret data in the context of the Wright–

Figure 3.4 Chinook salmon (*Oncorhynchus tshawytscha*). The effective population size of the Chinook salmon is heavily studied, because several Chinook salmon populations are endangered. Studies often estimate very small values of N_e for this species. This may in part be because the population size fluctuates between years, and in part because typically only few individuals leave many offspring while most individuals leave no offspring. This creates a high variance in offspring number that will reduce the effective population size.

Fisher model, or other similar models. For this purpose, the concept of **effective population size** has been invented. The effective population size of a population is the number of individuals in a Wright–Fisher model, or some similar model, that would produce the same amount of genetic drift as in the real population. The amount of genetic drift can be measured in various ways—for example, as the expected heterozygosity; as the expected number of average pairwise differences; or simply as the rate of coalescence, which, as we have argued, predicts measures of variability such as the average pairwise difference. If we say that a population has an effective population size of 200 with respect to heterozygosity, it means that the population harbors the same amount of heterozygosity as a Wright–Fisher population of 200 individuals. It is important to realize that the true number of individuals in the population can be vastly different from its effective population size (**Figure 3.4**).

Several factors can affect effective population size. One is, obviously, a change in the real population size. Sometimes we can approximate a population of changing size in terms of an idealized population of constant size. Imagine, for example, that a population fluctuates from generation to generation between $2N$ and $2Na$ individuals, where a can be any positive number. Then the probability of coalescence of a pair of lineages is $1/(2N)$ half of the time and $1/(2Na)$ half of the time. On average, the probability of coalescence is thus $0.5 \times 1/(2N) + 0.5 \times 1/(2Na)$. As the rate of coalescence is equal to 1 divided by the population size, a Wright–Fisher population with a corresponding rate of coalescence would have population size of $[0.5 \times 1/(2N) + 0.5 \times 1/(2Na)]^{-1}$. Or we could write $2N_e = [0.5 \times 1/(2N) + 0.5 \times 1/(2Na)]^{-1}$. More generally, if a population fluctuates between population sizes $N_1, N_2, \ldots, N_k$ at a proportion $p_1, p_2, \ldots, p_k$ of the time, respectively, a Wright–Fisher population with the same rate of coalescence would have an effective population size (fluctuating population size) of:

Harmonic Mean

$$N_e = \frac{1}{p_1/N + p_2/N_2 + \ldots p_k/N_k}$$ (3.8)

effective pop. size for a fluctuating pop.

This is the **harmonic mean** of the population size. So in the case of rapidly fluctuating population sizes, the **coalescent effective population size** is the harmonic mean of the real population size. The harmonic mean differs from the traditional arithmetic mean (average) by giving smaller population sizes more weight than larger population sizes. So the harmonic mean will be smaller than the arithmetic mean. The use of this concept of effective population size works well only when population sizes change quite rapidly. If population sizes are relatively stable, but the population has experienced a few strong changes in population size, such as a single bottleneck sometime in the past, the harmonic mean of the population size will not accurately predict the amount of genetic variability in the population.

In species with two sexes, differences in the number of males and females (unequal sex ratio) can also influence the effective population size. Assume there are N_f females and N_m males, so that the total population size is $N_f + N_m$. Because everybody receives half of their alleles from their mother and half from their father, the chance that two alleles sampled at random from a large population have both been transmitted maternally in the previous generation is $\frac{1}{2} \times \frac{1}{2} = \frac{1}{4}$. If they have both been maternally transmitted, the probability that they are from the same female in the previous generation is $1/N_f$. The probability that this female transmitted the same allele to both offspring is $\frac{1}{2}$, so the total probability of coalescence in a female in the previous generation is $1/(8N_f)$. Using similar arguments for males, we find that the total rate of coalescence per generation is $1/(8N_f) + 1/(8N_m)$. For an idealized Wright–Fisher population, the rate of coalescence is $1/(2N_e)$; setting that equal to the expression for a population with two sexes, we find that the effective population size (unequal sex ratio) is:

$$N_e = \frac{1}{(4N_m)^{-1} + (4N_f)^{-1}} = \frac{4N_m N_f}{N_m + N_f}$$ (3.9)

An unequal sex ratio will reduce the effective population size below $N_f + N_m$. The more unequal the sex ratio is, the stronger the reduction.

Another important factor that can influence effective population size is variance in offspring number. If some individuals have many more offspring than expected under a Wright–Fisher model, and other individuals have many fewer, the population behaves as if it were smaller than an idealized Wright–Fisher population with the same number of individuals. Like the case of fluctuating population sizes, or differences in sex ratio, the calculation of effective population size can be modified to take this into account.

While there are other ways of defining effective population sizes, most correspond closely to the concept of the coalescent effective population size.

Interpreting Estimates of θ

θ has a special role in population genetics theory because it determines, at least in the simplest of models, the amount of genetic variability observed in the population. As an example, consider the data by Wall et al. (2008) in **Table 3.1**. Wall surveyed variation in DNA sequences of X chromosomes and autosomes in a number of human populations. The survey included three African population groups: Mandenka, Biaka, and San; one Asian population group: Han; one European population group: Basque; and a Melanesian population group. They used Tajima's estimator to estimate θ (π in Table 3.1). θ can be reported for the entire sequence, or per site (the value for the entire sequence divided by the number of sites). Wall et al. calculated it per site, and reported it as a percentage.

Notice that for the African groups, the estimate of θ in autosomes varies between 0.0012 and 0.00126 per site. For the other population groups, the value is smaller, between 0.00078 and 0.00087. Why is θ larger in African population groups than in other population groups? As $\theta = 4N\mu$, two factors influence θ, the population size and the mutation rate. It is unlikely that the mutation rate is systematically different between different population groups. So the most likely explanation is that the effective population sizes are larger in Africa than in other areas. This might seem strange, as the number of Biakas or San individuals is vastly fewer than the number of, say, Han individuals alive today. There are more than 1 billion Han Chinese individuals in the world, but only about 30,000 Biakas. However, the simple model we have entertained here, assuming a single randomly mating population of constant size, is not a very realistic model for human populations. There is considerable population subdivision, especially in Africa, leading to deviations from the random mating assumption. Moreover, human populations have undergone many changes in size through time; for example, the Han population has expanded drastically over

TABLE 3.1
Averages of basic summary statistics for six population samples

	π (%)
Autosomes	
Mandenka	0.120
Biaka	0.121
San	0.126
Han	0.081
Basque	0.087
Melanesians	0.078
X chromosomes	
Mandenka	0.099
Biaka	0.095
San	0.085
Han	0.058
Basque	0.071
Melanesians	0.066

Source: Wall et al. (2008).

the past 4000–5000 years. Finally, humans are thought to have originated in Africa and have subsequently spread to other continents. As part of that process, much of the original genetic variability might have been lost from the populations outside Africa, but maintained inside Africa. For all of these reasons, effective population sizes in humans have very little to do with actual population sizes. So when we say that the effective population size of the African populations is larger than those of the other populations, it should not be interpreted as meaning that there are more individuals in these populations. Rather, these populations have been able to maintain more genetic variability for other reasons.

We also notice from the table that θ in general is larger in the autosomes than in the X chromosomes. This could be due to a difference in mutation rate, but there is a more likely explanation. Women carry two copies of the X chromosome, but men carry only one. If there is an equal sex ratio (equal numbers of men and women), there are therefore only three copies of an X chromosomal gene for every four copies of an autosomal gene. The effective population size of the X chromosomes is, therefore, only three-quarters that of the autosomes.

The Infinite Alleles Model and Expected Heterozygosity

Another population genetic model of mutation is the **infinite alleles model** (sometimes also called the infinitely many alleles model). The infinite alleles model assumes that every new mutation generates a new allele. If we again imagine the population as a bag of marbles, different alleles are represented by marbles of different color. Every mutation has the effect of changing the color of a marble into a new color, not previously observed in the population. As each new mutation results in a new, previously unobserved allele, there are, in principle, infinitely many possible different alleles in this model. This model may be appropriate if we categorize the data in terms of different allelic types, as was done with alleles distinguished by protein electrophoresis, and if we are not interested in keeping track on how these types relate to each other, i.e., we are just keeping track of whether the DNA is different or identical between different gene copies.

Based on this model and the coalescence for $n = 2$, we can calculate the probability that two sampled gene copies are identical to each other. That is simply the probability that no mutations happened on the two ancestral lineages between the time of coalescence and the present. We have previously argued that the rate at which mutations occur at each of the two lineages is $\theta/2$. We also know that the mean time to coalescence for two lineages is 1, and that this time is exponentially distributed. This implies that the rate at which coalescences occur, when looking backward in time, is 1. A result from probability theory (see last paragraph of Appendix B) then tells us that the chance that a coalescence occurs before the first mutation (again looking backward in time) is the rate at which coalescence events occur divided by sum of the rate of mutation and the rate of coalescence, i.e.,

$$\text{Pr(}coalescence\ before\ mutation) = \frac{1}{1+2\theta/2} = \frac{1}{1+\theta} \tag{3.10}$$

[handwritten: chance 2 gene copies all identical]

So the chance that two gene copies at a locus are identical is simply $1/(1+\theta)$. Under random mating, the probability that two gene copies are identical is the same as the expected homozygosity. Likewise, the probability that two gene copies are different from each other is the expected heterozygosity. As the sum of the expected heterozygosity and expected homozygosity must equal 1, we have

$$E[heterozygosity] = 1 - \frac{1}{1+\theta} = \frac{\theta}{1+\theta} \tag{3.11}$$

So if we sample an individual from a randomly mating population, we would expect the proportion of loci for which the individual is heterozygous to be $\theta/(1+\theta)$, if the infinite alleles model and the standard coalescence model are true. This result would be the same for the infinite sites model. In that case, we would interpret the heterozygosity as the haplotype heterozygosity, i.e., the probability that the two haplotypes are nonidentical.

[handwritten: assume infinite alleles model & std. coalescence]

The probability that two gene copies (DNA sequences) are identical is the same as the probability that no mutations happen before the coalescence event, which again is $1/(1+\theta)$. However, if there are back-mutations, with allele A changing to allele a and then back again, then two gene copies can be identical, not because no mutations have happened, but because one mutation has canceled the effect of another mutation. Population geneticists like to distinguish between Identity by State (IBS) and Identity by Descent (IBD). IBS is simply identity of observed alleles, and IBD is identity due to an absence of mutation. In the infinite alleles model and the infinite sites model IBS and IBD are identical; all identity is caused by an absence of mutation. However, in models with back-mutation, IBD and IBS are not the same thing. For example, a number of relatively rare disease alleles occur repeatedly, de novo, by mutation. One case is the 3034del4 mutation in the *BRCA2* gene. This mutation predisposes the carriers to early-onset breast cancer, and occurs in mutation-prone regions of *BRCA2*. Individuals may carry this allele because one of their parents also carried it and transmitted it to them. Alternatively, they may carry the mutation because it arose as a de novo mutation during meiosis, as demonstrated for a particular patient in a study by Van der Luijt et al. (2001). In the latter case, the new mutation would be IBS, but not IBD, with the other 3034del4 alleles already segregating in the population.

[handwritten: IBS vs. IBD]

The Coalescence Process in a Sample of *n* Individuals

We have found a mathematical description of the coalescence process for two gene copies. We can make a similar description for a sample of size *n* from a population of size 2*N*. The structure of the lineages will then form a tree, where each point in which lineages meet (nodes) corresponds to a coalescence event—a merging of lineages (**Figure 3.5**).

For the case of two gene copies, we saw that the rate of coalescence was 1 (when scaling time by 2*N*). We have also discussed that when there are *n* different objects, there are $n(n-1)/2$ distinct pairs of objects. So when there are $n(n-1)/2$ pairs of lineages, and each pair of lineages coalesce at rate 1, it should make sense that the total rate of coalescence is $n(n-1)/2$. That *for sample of* turns out, in fact, to be true. A major accomplishment of J. F. C. Kingman *n gene copies* was to show that when considering the limit of large population sizes (*N* *(lineages)* → ∞), a coalescence process arises in which we can ignore the possibility that three or more lineages could coalesce at the same time. Thanks to Kingman, we can simply piece together the coalescence process for many gene copies (*n* > 2) from the coalescence process for a single pair of gene copies

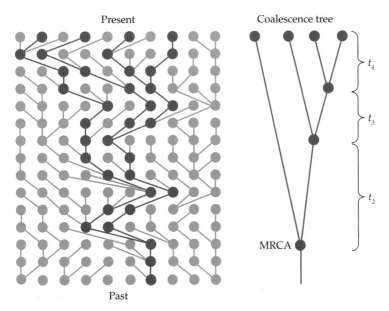

Figure 3.5 The ancestry of a sample of four individuals. The interpretation of the figure is similar to the one provided for Figure 3.2. t_4, t_3, and t_2 are the times in the coalescence tree during which there are 4, 3, and 2 lineages, respectively.

($n = 2$). Again, we are looking backward in time and waiting for the first coalescence event to happen. We previously used the analogy of waiting for a taxi in Manhattan. Now imagine that there are $n(n-1)/2$ taxi companies, and we are waiting for the first empty taxi to drive by from any of these $n(n-1)/2$ companies. The first taxi represent the first coalescence event among any of the $n(n-1)/2$ possible coalescence events. The time until this happens is again exponentially distributed, but the rate is now $n(n-1)/2$. The mean time we have to wait until the first coalescence event is 1 divided by the rate, or $1/[n(n-1)/2] = 2/[n(n-1)]$. After the first coalescence event, there are $n-1$ lineages left, and the process starts over with $n-1$ instead of n lineages The additional time (looking backward) we have to wait until the first coalescence event among these $n-1$ lineages is then exponentially distributed with mean $2/[(n-1)(n-2)]$, and so forth, until there is only one lineage left: the MRCA for all individuals has been found.

The Coalescence Tree and the tMRCA

The time it takes to go from k to $k-1$ lineages in the coalescence tree is t_k (Figure 3.5). The most important point to remember from the previous section is that $E[t_k] = 2/[k(k-1)]$ for all values of k, $2 \leq k \leq n$. For example, when there are four lineages, the mean time it takes before the first coalescence event occurs, looking backward in time, is $E[t_4] = 2/(4 \times 3) = \frac{1}{6}$. When there are three lineages, the expected time until the first coalescent even is $E[t_3] = 2/(3 \times 2) = \frac{1}{3}$, and when there are two lineages, the expected coalescence time is $E[t_2] = 2/(2 \times 1) = 1$, as before. So the closer you are to the present (the top of the tree in Figures 3.2 and 3.6), the shorter the coalescence times. A considerable amount of time in the tree is spent waiting for the very last coalescence event between the two last lineages. The standard neutral coalescence process generates trees with short **external lineages**. An external lineage is a lineage leading to a **leaf node** or simply a **leaf**. A leaf is a node in the tree representing one of the sampled individuals. Nodes further down in the tree are called **internal nodes**, and lineages connecting internal nodes are called **internal lineages**. The expected shape of a tree produced by the standard coalescence process is one in which the deep internal lineages are relatively long and the external lineages are relatively short. The node representing the MRCA is also called the **root** of the tree (**Figure 3.6**).

Armed with these results, we can immediately derive a number of important population genetic results. For example, what is the expected time to the most recent common ancestor (E[tMRCA]) for n haploid individuals? We have just shown that the expected time in the tree for which there are k lineages is $E[t_k] = 2/(k(k-1))$, so the expected tMRCA can be found by summing over all intervals in the tree:

$$E[tMRCA] = \sum_{k=2}^{n} E[t_k] = \sum_{k=2}^{n} \frac{2}{k(k-1)} \tag{3.12}$$

For example, in a sample of five individuals, the expected time to their most recent ancestor is $2/2 + 2/6 + 2/12 + 2/20 = 1.6$ ($\times 2N$ generations). If have had a sample of fifty individuals, the expected tMRCA would instead be 1.96–only slightly larger than the expected tMRCA for a sample of five individuals. And even a very large (infinite) sample would not have an expected tMRCA larger than 2. So a large sample will, on average, have just a slightly older MRCA than the small sample. This illustrates an important principle about the coalescence process. There is less and less additional information as more and more sequences are sampled. Much of the genetic variability can be captured by examining just a handful of individuals, because with five individuals, you are likely to already have sampled enough individuals to have the MRCA of the entire population spanned by the ancestral lineages of your sample (**Figure 3.7**).

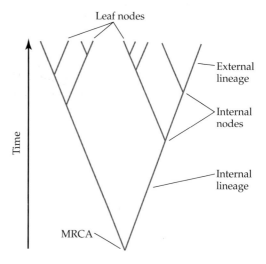

Figure 3.6 An example of a coalescence tree.

Total Tree Length and the Number of Segregating Sites

A quantity of critical importance is the number of segregating sites (S), i.e., the number of sites that are variable in a sample of DNA sequences from a population. For any particular DNA sequence data set, the value of S is found by simply counting the number of sites for which any on the DNA sequences carries a different allele. For example, for the data given in the Infinite Sites Model section, $S = 3$.

To derive the expected number of segregating sites, we first need to derive the **total tree length**, i.e., the sum of the length of all lineages in the coalescence tree. We can easily derive this using the previous results for the expected time in the tree for which there are k lineages, t_k:

$$E[total\ treelength] = \sum_{k=2}^{n} kE[t_k] = \sum_{k=2}^{n} \frac{2k}{k(k-1)} = 2\sum_{k=2}^{n} \frac{1}{k-1} = 2\sum_{k=1}^{n-1} \frac{1}{k} \quad (3.13)$$

Under the infinite sites model, the number of segregating sites is simply the total number of mutations that occur on any lineage. We know

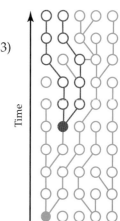

Figure 3.7 An illustration of the difference between the MRCA of a sample and the MRCA of a population. Each circle indicates a haploid individual, and each line is a line of descent indicating a parent–offspring relationship. In this hypothetical example, the population size is 5, and the sample size is 2. The sample and all its ancestors are shown in red. The MRCA of the sample and the MRCA of the population are shown in solid red and blue, respectively.

that the expected number of mutations on a lineage of length τ is $\tau\theta/2$. We are interested in mutations in any position of the tree, so τ now represents the total tree length. We find the expected number of segregating sites as

$$E[S] = \left(2\sum_{k=1}^{n-1}\frac{1}{k}\right)\frac{\theta}{2} = \theta\sum_{k=1}^{n-1}\frac{1}{k} \tag{3.14}$$

Expected # segregating sites

We have previously discussed how θ can be estimated from the average number of pairwise differences. Another possible method is to use S. We could rearrange the equation above to find

$$\theta = \frac{E[S]}{\sum_{k=1}^{n-1}\frac{1}{k}} \tag{3.15}$$

This might suggest the following estimator of θ named after the population geneticist and statistician G. A. Watterson (Watterson's estimator):

$$\hat{\theta}_w = \frac{S}{\sum_{k=1}^{n-1}\frac{1}{k}} \tag{3.16}$$

Notice that this estimator, like Tajima's estimator, is unbiased, as

$$E[\hat{\theta}_w] = E\left[S\bigg/\sum_{k=1}^{n-1}\frac{1}{k}\right] = E[S]\bigg/\sum_{k=1}^{n-1}\frac{1}{k} = \theta \tag{3.17}$$

As an example, consider again the data from the Infinite Sites Model section. For these data, $S = 3$ and $n = 5$, so we get

$$\hat{\theta}_w = \frac{3}{\sum_{k=1}^{5-1}\frac{1}{k}} = \frac{3}{1+\frac{1}{2}+\frac{1}{3}+\frac{1}{4}} = 1.44 \tag{3.18}$$

We previously obtained an estimate of $\hat{\theta}_T = 1.6$. The two estimates are similar but not identical. Both estimators are unbiased, so why do they give different results? It could occur by chance alone. The two estimators provide reasonable methods for guessing the true value of θ, but the guesses are not identical. If the estimates were guaranteed to be identical, the estimators would have to be the same. But there could also be something else going on. If the difference between the estimates is larger than we can explain by chance alone, it might suggest that there is some problem with the estimators or the models under which they have been derived. We will use this fact to our advantage in Chapter 9.

Calculating the full probability distribution of S is somewhat harder than finding the expectation, but we can do it relatively easily for a sample of size $n = 2$. We consider the coalescence of two genes and recall that the rate of coalescence is 1, and that the rate of mutations is θ when adding together the rates from both ancestral lineages. We have already argued that the chance that the first event (when looking back in time) is a mutation is $\theta/(1 + \theta)$ and that the chance that the first event is a coalescence event is $1/(1 + \theta)$. If the first event is a mutation event, the process starts over again, and looking back in time, coalescence and mutation events again happen at rates $1/(1 + \theta)$ and $\theta/(1 + \theta)$, respectively. Therefore, the chance of first observing one mutation event and then observing a coalescence event is $\theta/(1 + \theta) \times 1/(1 + \theta)$. This is the probability of exactly one mutation occurring in the coalescence history of the two gene copies, i.e., it is the probability that $S = 1$. Similarly, the probability of seeing any other number of segregating sites, S, is the probability that exactly S mutations occur on the ancestral lineages of the two gene copies before the first coalescence event, when looking back in time. We realize that this probability is ($n = 2$):

$$\Pr(S = j) = \frac{1}{1+\theta}\left(\frac{\theta}{1+\theta}\right)^{j}, j = 0, 1, 2, \ldots \qquad (3.19)$$

The Site Frequency Spectrum (SFS)

We have so far discussed two possible summaries of DNA sequence data: the number of segregating sites (S) and the average number of pairwise differences (π). These are only two possible summaries, and neither provides much information regarding allele frequencies. If we are interested in allele frequencies, an alternative summary is the **site frequency spectrum** (**SFS**). The site frequency spectrum is obtained by tabulating the sample allele frequencies of all mutations. As an example, consider again the data example in the Infinite Sites Model section. There are five sequences and three segregating sites. Assume for the sake of this example that any allele coded as 1 in the binary matrix is a *derived allele* and any allele coded as '0' is an *ancestral allele*. The **ancestral allele** is the allele found in the MRCA of the sample. A **derived allele** (a mutated allele) is an allele that is not ancestral. In the data example, the frequencies of the derived alleles in the sample are $2/5$, $2/5$ and $4/5$ (verify this yourself). In other words, the proportion of derived alleles with a frequency of $1/5$, $2/5$, $3/5$, and $4/5$ in the sample is $f_1 = 0, f_2 = 2/3, f_3 = 0$, and $f_4 = 1/3$, respectively, and we can write the SFS as a vector $f = (f_1, f_2, \ldots, f_{n-1})$ for a sample of n (haploid) individuals. We can plot the SFS in a histogram as in **Figure 3.8**.

The SFS discussed so far assumes that it is known which allele is ancestral and which is derived. However, this is generally not known from the sequence data itself. To determine which allele is derived and which is ancestral, researchers typically examine some **outgroups**—other closely related species. For example, if there is a C/T polymorphism in humans, and chimpanzees, gorillas and orangutans all have a C in this position, it is highly likely that the ancestral allele is C. If information is not avail-

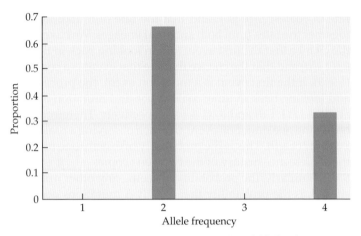

Figure 3.8 The site frequency spectrum (SFS) for the DNA sequence data example from the infinite sites model section.

able regarding closely related outgroups, one can also *fold* the frequency spectrum. The folded SFS contains no information regarding ancestral or derived states. The **folded frequency spectrum**, f^*, is obtained by adding together the frequencies of the derived and ancestral alleles, i.e., by setting $f^*_j = f_j + f_{n-j}$ for $j < n/2$ and $f^*_j = f_j$ for $j = n/2$. It is only defined for values of $f^*_j < n/2$. In our example from before, we have $f^*_1 = \frac{1}{3}, f^*_2 = \frac{2}{3}$, and $f^*_3 = 0$.

The SFS clearly includes much more information regarding the data than just S or π, because S and π can be calculated directly from f, but the opposite is not true. To make use of the SFS, we need to be able to calculate the expected SFS under the coalescence model. We can do that using our previously gained insights regarding coalescence trees.

Let us consider f_1, the proportion of derived alleles segregating at a frequency of $\frac{1}{n}$ in the sample. Such mutations are often called **singletons**. Under the infinite sites model, any mutation that lands on an external lineage in the coalescence tree results in a singleton, but no mutations landing on internal lineages do. Using considerations of the structure of the coalescence tree, it can easily be shown (this is left as an exercise) that the expected total length of external lineages in the tree is 2 (again when measuring time in units of 2N), independent of the sample size. As the expected number of mutations on any set of lineages of total length τ is $\tau\theta/2$, the expected number of singletons is, therefore simply θ. The expected total number of mutations in the tree is $\theta\sum_{k=1}^{n-1}\frac{1}{k}$, and the expected proportion of derived singletons, under the infinite sites model, is therefore

$$E[f_1] = \frac{\theta}{\theta\sum_{k=1}^{n-1}\frac{1}{k}} = \left[\sum_{k=1}^{n-1}\frac{1}{k}\right]^{-1} \tag{3.20}$$

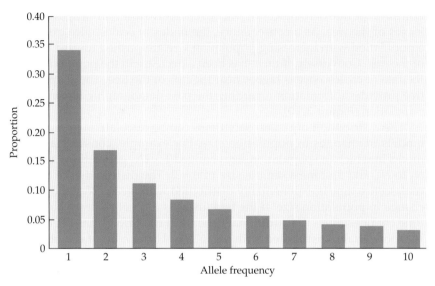

Figure 3.9 The expected site frequency spectrum (SFS) for a sample of $n = 10$ haploid individuals under the standard neutral coalescence model with infinite sites mutation.

Using similar logic for other categories of mutations, we can find the more general expression

$$E[f_j] = \frac{1/j}{\sum\limits_{k=1}^{n-1} \frac{1}{k}}, \, j = 1, 2, \ldots, n-1 \tag{3.21}$$

This expression provides us with the expected SFS under the standard coalescence model with infinite sites mutations. In **Figure 3.9**, notice that singletons are the most common class of mutations. Most mutations are of low frequency in the sample, and mutations of high frequency are relatively rare. We will discuss the SFS in more detail in the chapters on estimating demographic parameters and on detecting natural selection.

Tree Shape as a Function of Population Size

The mean time in the coalescence tree, measured in $2N$ generations, in which there are k lineages in the tree (t_k; Figure 3.5) is $2/[k(k-1)]$. Measured in terms of the number of generations, the expected time is $2N/[k(k-1)]$. The coalescence time is proportional to the population size; lineages in small populations coalesce quickly, and lineages in large populations coalesce slowly. However, if the population size changes over time, so will the rate of coalescence. For example, if there has been a strong increase in population size, the rate of coalescence will be relatively slow for lineages near the leaves of the tree. As a consequence, the nodes in coalescence trees from a population with increasing population size are pushed toward the root of the tree. Conversely, if there has been a decrease in the population size, the rate of coalescence is higher near the tips of the tree. The shape of the tree provides information regarding the past demographic history of the population (**Figure 3.10**).

The distribution of coalescence times can be calculated under various models of changes in population size. For example, consider a simple model in which the population size changed $T \times 2N_2$ generations ago from N_1 to N_2, and assume a sample size of $n = 2$. The distribution to the first coalescence time is exponentially distributed with mean $2N_2$, ignoring the change in population size. If the two lineages did not coalesce before the time changes, then the additional time until the MRCA has been found is exponentially distributed with mean $2N_1$. Using a bit of calculus (and material from Appendix B), we then find

$$E[t] = 2N_1 e^{-T} + 2N_2 \left(1 - e^{-T}\right) \tag{3.22}$$

If T is very small, i.e., the change in population size happened very recently, the expected coalescence time is determined mostly by the ancestral population size (N_1). If the time change happened a long time in the past, i.e., T is large, then the expected coalescence time is mostly determined by N_2, the current population size. Similar calculations can be done for other models of population size change, and for larger sample sizes, to predict the effect of changing population sizes on genetic variability.

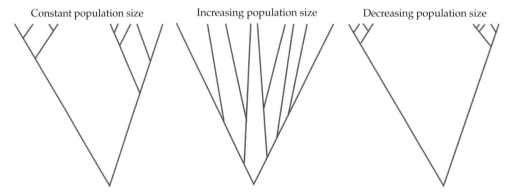

Constant population size   Increasing population size   Decreasing population size

Figure 3.10 Three coalescence trees representing a population of constant size, a population with increasing size, and a population with decreasing size.

References

*Donnelly P. and Tavaré S., 1995. Coalescents and genealogical structure under neutrality. *Annual Review of Genetics* 29: 401–421.

*Hudson R. R., 1991. Gene genealogies and the coalescent process. *Oxford Surveys in Evolutionary Biology* 7: 1–44.

*Kingman J. F. C., 1982b The coalescent. *Stochastic Process and their Applications.* 13: 235–248.

*Tajima F., 1983. Evolutionary relationship of DNA sequences in finite populations. *Genetics* 105: 437–460.

Van der Luijt R. B., van Zon P. H. A., Jansen R. P. M., et al., 2001. De novo recurrent germline mutation of the *BRCA2* gene in a patient with early onset breast cancer. *Journal of Medical Genetics* 38: 102–105.

Wall J. D., Cox M.P., Mendez F.L., et al., 2008. A novel DNA sequence database for analyzing human demographic history. *Genome Research* 18: 1354–1361.

*Wang J., 2005. Estimation of effective population sizes from data on genetic markers. *Philosophical Transactions of the Royal Society of London B: Biological Sciences* 360: 1395–1409.

*Recommended reading

EXERCISES

3.1 In a sample of five gene copies, what is the expected time to the most recent common ancestor under the standard coalescence model measured in *2N* generations? What is the expected total tree length?

3.2 A researcher sequences a 10-kb-long DNA sequence from a single individual. The mutation rate in the region is 10^{-9} per site. The researcher finds that 21 of the sites are heterozygous. Assuming an infinite sites model and a standard neutral coalescence model, provide an estimate of the effective population size of the population from which this individual has been sampled.

3.3 A researcher sequences 5 diploid individuals (10 DNA sequences) from a population with an effective population size of 20,000 individuals. The total mutation rate for the region is 10^{-5} per generation. Assuming a standard coalescence model and infinite sites mutation, how many segregating sites should the researcher expect to find in the data?

3.4 For the data and assumptions in Exercise 3.3, what is the expected haplotype homozygosity?

3.5 The following DNA sequence data were obtained from a single population:

AAGATGACAGATAGGCA

CTGGTGACTGATAGGCA

CTGGTGACTGATAGGCT

CAGATGACTGATAGGCT

How many segregating sites are there? What is the average number of pairwise differences (π)?

3.6 Using the data from exercise 3.5, calculate two estimates of θ, one based on Watterson's estimator and one based on Tajima's estimator.

3.7 Make a histogram of the site frequency spectrum (SFS) for the data in Exercise 3.5.

3.8 Do the data in Exercise 3.5 contain more, fewer, or the same number of singletons as expected under the standard neutral coalescence model?

3.9 A researcher obtains a sample of two diploid individuals (four gene copies) from a population for which the assumptions of the standard coalescence model holds true. What is the probability that the two sequences from the first individual are more closely related to each other than they are to any of the sequences from the second individual, i.e., that they share a most recent common ancestor more recently with each other than with any of the other two sequences?

3.10 Prove by induction that the total length of external lineages in the tree is 2 ($\times$ 2N) in the standard coalescence model, by first arguing that the result is true for a sample size of $n = 2$ and then showing that if it is true for a sample size of $n - 1$, then it is also true for a sample size of n. Hint: If the expected total length of external lineages is 2 in a tree with $n - 1$ leaves, and there are $n - 1$ external lineages, then the expected length of each external lineage is $2/(n - 1)$.

4 *Population Subdivision*

WE HAVE SO FAR CONSIDERED models of a single population with random mating. However, real populations often have some structure, typically geographically determined, in which individuals living close to one another are more likely to mate than more geographically distant individuals. When the population is not randomly mating because of geographic structure, we say that there is **population subdivision** or **population structure**. Many of the most important applications of population genetic theory involve making inferences about geographic and historical patterns of population subdivision. Population subdivision is important for understanding evolution and the effects of genetic drift and natural selection, and it is also often of direct importance in conservation biology, for the management of rare or endangered species. In a number of different organisms, ranging from brown bears in Scandinavia to sockeye salmon in Alaska and black rhinos in South Africa, researchers have used genetic markers to determine which groups of individuals should be considered separate genetic units. Policy decisions on management of the species depend on this type of classification on individuals.

This chapter will focus on the effects of population structure of genetic differentiation, and on methods for quantifying such differentiation.

The Wahlund Effect

Many different population genetic models take population subdivision into account. We will start by examining the simplest possible model: two subpopulations, each of which is in HW equilibrium. Assume a di-allelic locus with alleles A and a, in which the frequencies of A are f_{A1} and f_{A2} in populations 1 and 2, respectively. The average frequency of allele A, when pooling the two populations, is

$$f_A = \frac{2N_1 f_{A1} + 2N_2 f_{A2}}{2N_1 + 2N_2} \tag{4.1}$$

where N_1 and N_2 are the sizes of populations 1 and 2, respectively. If the populations have the same sizes, we have simply

$$f_A = \frac{f_{A1} + f_{A2}}{2} \tag{4.2}$$

If each subpopulation is in HW equilibrium and the population sizes are the same, the proportion of heterozygous individuals is

$$H_S = \frac{2f_{A1}(1-f_{A1}) + 2f_{A2}(1-f_{A2})}{2} = f_{A1}(1-f_{A1}) + f_{A2}(1-f_{A2}) \tag{4.3}$$

We use the subscript S to indicate that this is the heterozygosity in the subdivided population. It represents the heterozygosity we would expect to observe if we went into the field, sampled both populations, and estimated the heterozygosity. However, the heterozygosity expected from a population with allele frequency $(f_{A1} + f_{A2})/2$ is

$$H_T = 2\frac{f_{A1}+f_{A2}}{2}\left(1 - \frac{f_{A1}+f_{A2}}{2}\right) \tag{4.4}$$

H_T represents the heterozygosity we would expect if the pooled population is in HW equilibrium. We use the subscript T to indicate that this is the value of H for the total (pooled) population.

Now let the difference between the allele frequencies in the two populations be $\delta = |f_{A1} - f_{A2}|$. By adding and subtracting $\delta^2/2$ on the right-hand side of the equation above, and rearranging terms, we find (verify this yourself):

$$H_T = f_{A1}(1 - f_{A1}) + f_{A2}(1 - f_{A2}) + \delta^2/2 \tag{4.5}$$

From the last equation, we see that if the allele frequencies are the same in the two populations ($\delta = 0$), then $H_T = H_S$. Because both populations are in HW equilibrium, and the allele frequencies are the same in the two populations, the pooled population is also in HW equilibrium, and the proportion of heterozygotes is the same in the subpopulations as in the pooled population. In other words, there is no detectable population subdivision.

However, if the allele frequencies differ between the two populations ($\delta > 0$) then $H_T > H_S$—that is, the population contains fewer heterozygotic individuals than expected, given the pooled allele frequency. Population subdivision will in this way always lead to a reduction in heterozygosity and an increase in homozygosity compared to a randomly mating population with the same (total) allele frequency. This decrease in heterozygosity is called the **Wahlund effect**. It is a quite general result that also holds true, for example, if the populations have different sizes, if there are more than two populations, and if the loci are multi-allelic and not di-allelic.

F_{ST}: Quantifying Population Subdivision

The most commonly used measure for quantifying population subdivision, i.e., differences in allele frequencies between populations, is Wright's F_{ST}.

Although other definitions are sometimes used, F_{ST} is commonly defined as the difference between H_T and H_S, standardized by H_T:

$$F_{ST} = \frac{H_T - H_S}{H_T} \qquad (4.6)$$

We previously defined H_T and H_S for a set of two populations, but we could also make similar definitions for more than two populations. H_T would then be calculated based on the pooled allele frequency from all populations, and H_S would be found by averaging the heterozygosity of all populations.

If the allele frequency is the same in all populations, then $H_T = H_S$ and $F_{ST} = 0$. The allele frequencies are maximally different when different alleles are fixed in different populations. In the di-allelic case and two populations, this situation would occur when $f_{A1} = 0$ and $f_{A2} = 1$ or $f_{A1} = 1$ and $f_{A2} = 0$. Then $H_T > 0$ and $H_S = 0$, so $F_{ST} = 1$. F_{ST} therefore varies between 0 and 1, with 0 implying no detectable population subdivision and 1 implying maximal amounts of differentiation. S. Wright, who first defined F_{ST}, provided some guidance to the interpretation of F_{ST}, suggesting that values between 0 and 0.05 indicate little or no differentiation, values between 0.05 and 0.15 indicate moderate differentiation, values between 0.15 and 0.25 indicate great differentiation, and values larger than 0.25 indicate very great differentiation. Values of F_{ST} vary greatly between different species, and also between different population comparisons within a species. For example, a comparison of the most highly differentiated populations of the North Atlantic humpback whale yields an F_{ST} value of about 0.04. This value is significantly different from zero, but still quite modest. In contrast, humpback whales in the Pacific (**Figure 4.1**) and in the Atlantic oceans have much stronger genetic differentiation from each other; comparisons yield F_{ST} values larger than 0.4.

Figure 4.1 The humpback whale (*Megaptera novaeangliae*) is an example of a species where extensive genetic work has been done to understand population structure in order to improve management of the species.

TABLE 4.1 Allele frequencies of disease-causing SNPs in two genes

Allele	Africa	Europe	East Asia
FTO rs9939609	0.471	0.426	0.157
TCF7L2 rs7901695	0.629	0.325	0.044

rs9939609 is a specific SNP in the *FTO* gene with an allele conferring individuals carrying the allele an increased risk of obesity. rs7901695 is a SNP in the *TCF7L2* with an allele conferring an increased risk of type 2 diabetes. Allele frequencies are given for three different continental groups. *Source:* Myles et al. (2008).

Notice the similarity between the definition of F in Equation 1.8 (from Chapter 1) and the definition of F_{ST} in Equation 4.6. In Equation 1.8 we defined F as a scaled difference between the expected and the observed heterozygosity. We could also interpret it as the relative decrease in heterozygosity due to deviations from HWE within a population. F_{ST} is defined similarly, but instead of comparing the expected heterozygosity under HWE within a population to that observed among individuals in the population, we compare the heterozygosity for a combined population to that observed within populations. F_{ST} now represents the relative reduction in heterozygosity due to population subdivision. Wright analyzed the connection between F, F_{ST} and other similar measures (called **F-statistics**) in an elegant theory that explains the relative contribution of inbreeding and population subdivision in explaining deviations from HWE.

There are several ways of estimating F_{ST} from data. In the case of a single di-allelic locus, F_{ST} can be estimated simply by calculating allele frequencies and substituting these sample allele frequencies into the formulas given above. When there are many loci and information must be combined among multiple loci, more accurate methods for estimating F_{ST} can be employed.

As an example, consider the data in **Table 4.1**. Considering first the SNP in the *FTO* gene in Africans and Europeans, we find that $f = (0.471 + 0.426)/2 \approx 0.449$ and $H_T = 2 \times 0.449 \times (1 - 0.449) \approx 0.495$. We also find $H_S = [2 \times 0.471 \times (1 - 0.471) + 2 \times 0.426 \times (1 - 0.426)]/2 \approx 0.494$. So $F_{ST} \approx (0.495 - 0.494)/0.495 \approx 0.002$. There is essentially no evidence of a difference in allele frequencies. Had we instead done the calculations for the SNP in the *TCF7L2* gene, we would have found an F_{ST} value of approximately 0.09. We can also calculate F_{ST} for all three population groups combined. For the *FTO* SNP, we would find that $f = (0.471 + 0.426 + 0.157)/3 \approx 0.351$ and $H_T = 2 \times 0.3515 \times (1 - 0.351) \approx 0.456$. $H_S = [2 \times 0.471 \times (1 - 0.471) + 2 \times 0.426 \times (1 - 0.426) + 2 \times 0.157 \times (1 - 0.157)]/3 \approx 0.417$. So $F_{ST} \approx (0.456 - 0.417)/0.456 \approx 0.084$.

Because the allele frequencies for different loci in the genome are variable, and because genetic drift might have had different effects at different loci, F_{ST} varies among loci in the genome. It is generally assumed that the average value of F_{ST} in a genome, calculated between one or more populations, is a good overall measure of the degree of genetic differentiation between the groups. However, very extreme values of F_{ST} in a few loci, as compared to most other loci in the genome, may indicate that processes other than genetic drift and mutation are affecting allele frequencies. We will return to this point when discussing methods for detecting the effects of natural selection in the genome.

The average value of F_{ST} varies between species. In most mammals, the value of F_{ST} when comparing different geographically separated populations varies between 0.1 and 0.8. Only rarely are populations with F_{ST} values much larger than 0.8 categorized by biologists as belonging to the same species. In humans, F_{ST} values vary depending on which populations are compared: between different European groups, F_{ST} varies from 0 to 0.025; between different continental groups, such as Asians, Africans, and Europeans, the average F_{ST} varies from 0.05 to 0.2. These values are much smaller than for other species, especially considering the large geographic range of humans.

The Wright–Fisher Model with Migration

F_{ST} describes the extent of variation in allele frequencies among populations. How does F_{ST} change with migration among populations? To answer this, we have to generalize models of single populations to allow for migration among them. We start with the Wright–Fisher model and consider a model with two populations. Consider, for example, the aforementioned example of humpback whales in the Pacific and the Atlantic oceans. The whales are not moving back and forth between the two oceans very much—but occasionally a migrant will swim from one ocean to the other. We will make the simplifying assumption that each population evolves as a Wright–Fisher model, but occasionally an individual from one population is replaced with an individual from the other. This occurs with probability $m_{1\rightarrow2}$ from population 1 to population 2, and $m_{2\rightarrow1}$ from population 2 to population 1. $m_{1\rightarrow2}$ and $m_{2\rightarrow1}$ are the **migration rates** per individual per generation (**Figure 4.2**).

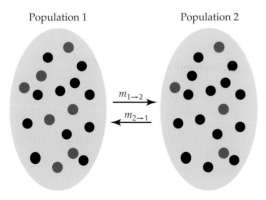

Population 1   Population 2

$m_{1\rightarrow2}$
$m_{2\rightarrow1}$

Figure 4.2 Two populations sharing migrants between them. If no other forces are affecting allele frequencies, eventually both populations will have the same allele frequencies.

We will further assume that the allele frequencies in the two populations are $f_{A1}(t)$ and $f_{A2}(t)$ at a particular locus in generation t. We can then find the expected allele frequency in population 1 in generation $t + 1$ as

$$E[f_{A1}(t + 1)] = (1 - m_{2 \to 1})f_{A1}(t) + m_{2 \to 1}\, f_{A2}(t) \qquad (4.7)$$

Similarly, the expected frequency in population 2 in generation $t + 1$ is

$$E[f_{A2}(t + 1)] = (1 - m_{1 \to 2})f_{A2}(t) + m_{1 \to 2}\, f_{A1}(t) \qquad (4.8)$$

A special case is one-way migration, which happens, for example, when a small island receives migrants from the mainland. For this type of model we again have $E[f_{A1}(t + 1)] = (1 - m_{2 \to 1})f_{A1}(t) + m_{2 \to 1}f_{A2}(t)$, where population 1 is the island population. Assuming that f_{A2} is not affected by migration and that no other forces are affecting allele frequencies, we let $f_{A2}(t) = f_{A2}$, i.e., it does not change through time. We can then find the equilibrium allele frequency for population 1 by setting $E[f_{A1}(t + 1)] = f_{A1}(t) = f_{A1}$; solving the equation for f_{A1}, we find that the allele frequency on the island will simply be $f_{A1} = f_{A2}$. Not surprisingly, if no other factors affect allele frequencies, the island population will eventually be expected to have the same allele frequency as the mainland population. In general, populations exchanging a lot of migrants will tend to eventually have similar allele frequencies. However, this effect is counteracted in each population by mutation, genetic drift, and possibly selection.

The Coalescence Process with Migration

We can extend the coalescence process derived in Chapter 3 to the case of two populations sharing migrants. We will track the ancestry of a sampled gene copy back in time, while keeping track of which population the ancestor of the individual belonged to (**Figure 4.3**).

   We use the two-population model described in the previous section, and to simplify a bit, we assume that the rate of migration per individual per generation is the same for both populations, i.e., $m_{1 \to 2} = m_{2 \to 1} = m$ (symmetric migration). This means that the probability of an individual in population 1 in generation t being a descendent of an individual in population 2 in generation $t - 1$ is m. We first consider the ancestry of a single gene copy from population 1. We have

$$\Pr(\textit{individual was not a migrant last generation}) = 1 - m \qquad (4.9)$$

Therefore,

$$\Pr(\textit{individual was not a migrant in the past r generations}) = (1 - m)^r \quad (4.10)$$

As before, we change the scaling of time and start measuring time in terms of $2N$ generations, and consider the limit of large populations. We set $r = 2Nt$, $M = 2Nm$ and let $N \to \infty$ to obtain

$$[1 - M/(2N)]^{2Nt} \to e^{-Mt} \qquad (4.11)$$

Population 1   Population 2

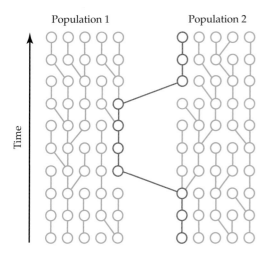

Time

Figure 4.3 The ancestry of a gene copy (red) from population 2 traced back in time in a two-population Wright–Fisher model. Dots represent individuals and lines represent parent–offspring relationships.

This implies that for sufficiently large populations, the time to the first migration event (looking back in time) is exponentially distributed (Appendix B) with expectation $1/M$.

Notice that this derivation is essentially identical to the derivation we did for the distribution of the coalescence time for a sample of two individuals ($n = 2$). We make the same assumption as before, that $N \to \infty$, but because we now set $M = 2Nm$, we will also assume $m \to 0$ for any particular value of M. The coalescence process with migration assumes small values of m (but not necessarily M) and large values of N.

Most of the time, for convenience, we will use the parameter M instead of m. In coalescence theory, it is more natural to use the parameters scaled by the population size. However, the parameter M also has a special interpretation that makes it very biologically relevant. Since there are $2N$ individuals in the populations, and the probability of any individual being a migrant is m, the actual number of migrants each generation is $2Nm = M$. From population genetic data, we can often estimate M directly, but we typically cannot know the value of m without knowing $2N$.

To simplify further, we will make the additional assumption that the two populations have equal population sizes: $N_1 = N_2 = N$. Consider, then, the ancestry of a single gene copy. At any time, the ancestor is in either population 1 or population 2 (Figure 4.3). When a gene copy is in population 1, you would expect the gene copy to stay there for $1/M$ coalescence time units (looking back in time). Similarly, when a gene copy is in population 2, you would expect to wait $1/M$ time units, on average, until it migrates to population 1 (again, looking back in time). So we now have a coalescence process, where in addition to coalescence events, we also allow lineages (ancestors of gene copies) to migrate between populations.

Figure 4.4 An illustration of the coalescence process for two individuals sampled from population 2. The ancestry of the sample is traced back in time in red until an MRCA is found. The periods in the tree in which the two lineages are in the same population (*S*) and in different populations (*D*) are indicated on the right.

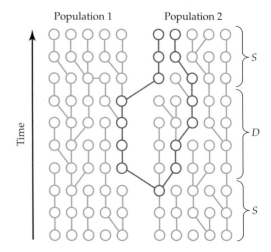

The coalescence process for two populations with migration is then similar to the standard coalescence process, but involves migration events occurring at rate M for each lineage. If there are i lineages in population 1 and j lineages in population 2, the total rate of migration of ancestral lineages on the coalescent time scale is then $(i + j)M$. At the same time, coalescence events occur in population 1 and population 2 at rates $i(i-1)/2$ and $j(j-1)/2$, respectively.

For example, for two individuals sampled from the same population, looking back in time, two different types of events may happen: either the two lineages coalesce, or one migrates into the other population (**Figure 4.4**). The first type of event occurs at rate 1 when time is scaled in terms of $2N$ generations, and the second type of event occurs at rate $2M$, because there are two lineages. If one of the lineages migrates, then the two lineages will be in different populations and cannot coalesce until one of the lineages has migrated into the other population. Coalescence events can only occur between two lineages within the same population. The reason for this is that we can ignore the possibility of a migration event and a coalescence event occurring in the same generation, as long as both $1/2(N)$ and m are small (we assume $m \to 0$ at the same rate as $2N \to \infty$).

Expected Coalescence Times for n = 2

The coalescence with migration model can be used to learn a lot about the genetics of a subdivided population. In the following example, we will derive the expected coalescence time for a sample size of two using the simple model given in the previous section for two populations of equal size. This is perhaps a bit more difficult than most of the other material in this book, but the derivation requires no special math skills, beyond knowledge of the properties of exponential random variables discussed in Appendix B.

Readers finding the material in Appendix B challenging may skip the proof and move directly to the discussion at the end of this section.

If you sample two gene copies, one from each of the two populations, how long (looking back in time) would you expect to have to wait to find a most recent common ancestor? To answer this question, we need to consider the possible ancestral configurations, i.e., we need to consider the separate cases of two lineages in the same population and two lineages in different populations. We will use the following notation: if the two lineages are in different populations, then the lineages are in configuration D (D for "different"; Figure 4.4). If the ancestor of both gene copies is in the same population, the lineages are in configuration S (S for "same"; Figure 4.4). We are interested in finding $E_S[t]$ and $E_D[t]$, the expected coalescence time for two lineages in the same population and in different populations, respectively. If the sample contains two individuals from different populations, then the process is initially (looking back in time) in configuration D, and there can be no coalescence event until one of the lineages has migrated. As previously discussed, for two lineages to coalesce, they must be in the same population. Each lineage migrates to the other population at rate M, so the total rate of migration is $2M$. This means that the expected time until one of them migrates is $1/(2M)$. If one of the lineages migrates, the configuration then changes from D to S, and we must wait an additional expected time of $E_S[t]$ until the coalescence event happens. We then have the following equation:

$$E_D[t] = \frac{1}{2M} + E_S[t] \tag{4.12}$$

Similar reasoning leads to

$$E_S[t] = \frac{1}{1+2M} + \frac{2M}{1+2M} E_D[t] \tag{4.13}$$

because in configuration S, the next event can either be a migration event, which occurs at rate M for each of the two lineages, or a coalescence event, which occurs at rate 1. The total rate is then $1 + 2M$, so the expected time we must wait (looking back) for the first event, either a migration or coalescence event, is $1/(1+2M)$. The probability that this event is a migration and not a coalescent is $2M/(1+2M)$, the rate of migration divided by the total rate of migration plus coalescence (see the last paragraph of Appendix B). If it was not a coalescent event but a migration event, we have to wait an additional time with expectation $E_D[t]$.

We now have two equations with two unknowns, $E_S[t]$ and $E_D[t]$. Solving this system of equations we find (two populations):

$$E_S[t] = 2 \text{ and } E_D[t] = \frac{1}{2M} + 2 \tag{4.14}$$

From this equation, we see that if M is small, the expected coalescence time for two individuals sampled from different populations is large. This should make intuitive sense: when M is small, it takes a long time until two ancestors of gene copies sampled from different populations are in the same population. If M is large, the coalescence time becomes relatively small and eventually becomes the same as the coalescence time for two individuals sampled from the same population. This should also make intuitive sense: if M is sufficiently large, the coalescence times are the same for individuals sampled within and between populations. What might be more surprising is that the expected coalescence time for two individuals sampled from the same population does not depend on M! No matter what the migration rate is, the expected coalescence time is just two on the coalescence time scale, equivalent to $4N$ generations on the natural time scale.

The results above can easily be extended to other models. For example, if there are d different populations, each of size $2N$ and with a symmetric migration rate m $[= M/(2N)]$ between all pairs of populations, a very similar calculation to the one given above shows that (d populations):

$$E_S[t] = d \text{ and } E_D[t] = \frac{1}{2M} + d \tag{4.15}$$

It is left as an exercise for the reader to verify that this result is correct. A model with d populations sharing migrants is often referred to as an **island model** with d populations. Sometimes the d populations in a model like this are instead called sub-populations or **demes**.

F_{ST} and Migration Rates

Assuming an infinite sites model, the expected number of pairwise differences can be calculated as twice the mutation rate, multiplied by the expected coalescence time (see Chapter 3). From the equations given in the previous section, we find that the expected number of pairwise differences between two sequences from the same population is $\mu \times 2 \times 2N \times 2 = 2\theta$. Similarly, the expected number of pairwise differences between sequences sampled from different populations is $\mu \times 2 \times 2N \times [1/(2M) + 2] = [1/(2M) + 2]\theta$.

We can use these results to predict the value of F_{ST} under the infinite sites model. Let k be the number of sites in the sequence. The infinite sites model describes the situation in which k is very large (goes to infinity), so that θ per site, θ/k, goes to zero. At this limit, the heterozygosity per site simply becomes the number of pairwise differences divided by the number of sites (because the chance that any two mutations hit the same site goes to zero). So we can write the expected heterozygosity per site as $2\theta/k$ for sequences sampled within a population, and $[1/(2M) + 2]\theta/k$ for sequences sampled from two different populations. We then have

$$H_S = 2\theta/k \tag{4.16}$$

and because there is equal probability, when sampling two sequences,

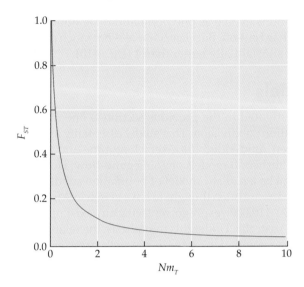

Figure 4.5 F_{ST} as a function of Nm_T in an island model in which there are N gene copies, and in which each population receives migrants at a rate of m_T per generation.

of sampling two from the same population and sampling one from each population, we have

$$H_T = (2\theta/k + [1/(2M) + 2]\theta/k)/2 = [1/(4M) + 2]\theta/k \quad (4.17)$$

F_{ST} is then given by (two populations):

$$F_{ST} = \frac{H_T - H_S}{H_T} = 1 - \frac{H_S}{H_T} = 1 - \frac{2\theta/k}{[1/(4M)+2]\theta/k} = 1 - \frac{2}{1/(4M)+2} = \frac{1}{1+8M} \quad (4.18)$$

Using similar lines of reasoning, this result can be extended to a number of other models, including models with more than two subpopulations, unequal migration rates between populations, etc. For example, using the equations for the island model with d populations given in the previous section, we find (d populations):

$$F_{ST} = \frac{(d-1)/d}{(d-1)/d+2dM} \quad (4.19)$$

The total number of migrants into a population is $m_T = (d - 1)m$, and as previously, $M = 2Nm$. If we consider a population divided into infinitely many subpopulations ($d \to \infty$), we find (infinitely many populations):

$$F_{ST} = \frac{1}{1+4Nm_T} \quad (4.20)$$

This famous result was obtained by S. Wright, using different techniques and somewhat different assumptions, many decades before coalescence theory was discovered.

The value of F_{ST} is shown as a function of Nm_T in **Figure 4.5**. Notice how fast the function decreases. Often, population geneticists argue that if

$Nm_T > 1$, then the populations evolve as one population. We notice from Figure 4.5 that this is not a hard cut-off. But clearly, there is not much population subdivision if Nm_T is substantially larger than 1 (e.g., >10).

Divergence Models

The model we have investigated in the previous section is one of many possible models of population subdivision. This model essentially assumes that populations have been subdivided for a very long time and that an equilibrium has been established. It models ongoing **gene-flow**. However, in many cases, this may not be a realistic model of population subdivision. For example, modern humans are often thought to have arisen in Africa and migrated out of there 60,000–110,000 years ago, replacing any existing hominids in Europe, Asia, and Australia. Under this hypothesis, the relationship between the different human population groups is not well characterized by an island model. To describe this type of population structure, we need to use **divergence models**. A divergence model is a model that describes populations diverging from common ancestral populations without subsequent gene-flow. As a hypothetical example, imagine that the humpback whales in the Atlantic Ocean originally came from the Pacific but migrated into the Atlantic thousands of years in the past, and that since then, no whales have been swimming back and forth between the Pacific and the Atlantic Oceans. We can then model the data by assuming that each population is currently evolving as an independent Wright–Fisher model that diverged from a common ancestral population sometime in the past. The process is illustrated in **Figure 4.6**, where we can think of population 1 representing Pacific whales and population 2 representing Atlantic whales. In the whale example, the initial divergence would have occurred when the first whales crossed from the Atlantic to the Pacific Ocean. Divergence models are similarly used to model human evolution, where one divergence event is the migration of the first humans out of Africa.

Different assumptions can be made regarding what happens at the time of divergence. For example, many models of human demography assume that there is a bottleneck in population size (a temporary strong reduction) at the time of divergence. For the sake of simplicity, we will here assume that there is no change in population size, and that the size of the ancestral population (N_A) equals that of both population 1 and population 2 ($N_A = N_1 = N_2 = N$). A coalescence process then arises where we follow the ancestry of the samples from each population independently for $2NT$ generations back in time, at which point we merge the two populations' ancestral samples. T is the divergence time between populations. After merging the two ancestral samples at the time of divergence, we can then further trace the ancestry of the sample until a final MRCA of the sample is found.

Notice that the divergence time between populations (T) is not the same as the coalescence time t. If we sample two sequences, one from each

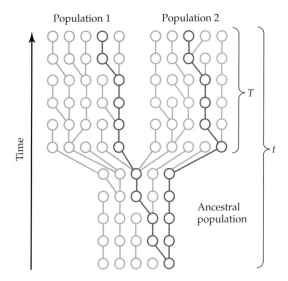

Figure 4.6 The coalescence process in a model with two diverging populations. At a time T in the past, the ancestral population split into two populations. The ancestry of a sample of two gene copies, one from each population, is traced back to the MRCA (red). The time to the MRCA (the coalescence time) is t. Notice that $t \geq T$.

population, we cannot estimate the divergence time by simply estimating the time of the MRCA of the two sequences. If we did so, we would tend to overestimate the divergence time. This is a very important point that often has been missed by researchers analyzing genetic data.

Expected Coalescence Times, Pairwise Difference and F_{ST} in Divergence Models

For the simple divergence model defined in the section above, it is fairly straightforward to calculate the expected coalescence time for two gene copies. First, two gene copies sampled from the same population behave exactly as two gene copies sampled under the standard neutral model, i.e., $E_S[t] = 1$ (when scaling time in terms of $2N$ generations). If we assumed N_A to be different from N_1 and N_2, we could similarly find the expected coalescence time simply as that of a standard population with a change in population size $2NT$ generations ago, as discussed in Chapter 3.

Two genes sampled from different populations cannot coalesce before the populations have merged. Looking back in time, after the populations have merged, the coalescence process proceeds just as in the standard coalescence model, so $E_D[t] = T + 1$. Differences in population sizes could also easily be incorporated in this case, but will be ignored here to keep the math as simple as possible.

Assuming an infinite sites model, the expected number of pairwise differences are then $\theta(T + 1)$ for sequences sampled from different populations and θ for sequences sampled from the same population. Using the same arguments as for the migration models, we also get $H_T = (T/2 + 1)\theta/k$ and $H_S = \theta/k$, and find

$$F_{ST} = 1 - \frac{H_S}{H_T} = 1 - \frac{\theta/k}{(T/2+1)\theta/k} = 1 - \frac{1}{T/2+1} = \frac{T}{T+2} \qquad (4.21)$$

When $T = 0$, there is no population subdivision, and $F_{ST} = 0$. As the divergence time becomes very large, F_{ST} approaches 1. The expression given above depends strongly on the assumptions regarding population size, and is perhaps less informative than the expression obtained for F_{ST} under the migration model. However, it is worth noticing that any particular value of F_{ST} can be explained by both a divergence model and a migration model. F_{ST} does not allow us to distinguish among different models of population history, and we do not learn anything about the plausibility of either model by simply estimating F_{ST}.

Isolation by Distance

In many species, the degree of population subdivision often increases with geographical distance. For example, González-Suárez et al. (2009) obtained DNA samples from California sea lions living in colonies in the Gulf of California and on the Pacific coast of Baja California, Mexico. **Figure 4.7** illustrates one of their findings. As the geographical distance increases between populations, so does F_{ST}. This is called **isolation by distance**. In fact, if you plot $F_{ST}/(1 - F_{ST})$ against geographic distance, there is approximately a linear relationship between the two.

How can we explain this result using the previously discussed population genetic models? We previously showed that under simplifying assumptions, $F_{ST} = 1/(1 + 8M)$ for two populations. So $M = (1 - F_{ST})/(8F_{ST})$. If the migration rate is a linear function of geographical distance, this would explain the result obtained by González-Suárez and colleagues.

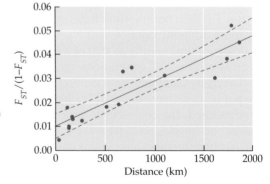

Figure 4.7 The relationship between $F_{ST}/(1 - F_{ST})$ and geographical distance (km) for California sea lion populations. Each dot represents a pair of populations, and a regression line is shown. (After González-Suárez et al., 2009.)

Another possibility is that migration occurs only between adjacent populations. Models based on this idea are called **stepping-stone models**. In general, they produce predictions similar to those that postulate gene-flow between all populations but assume that M is a function of distance between populations. Stepping-stone models are often used to understand patterns of isolation by distance.

However, divergence models may also explain the observed correlation between geographic distance and genetic differentiation observed in many species. Imagine a series of divergence events, each one of which results in one population splitting apart from another. Also assume that every time a divergence event occurs, the new population occupies an area adjacent to its parent population. Populations located close to each other will tend to be more genetically similar than distant populations, just as predicted in migration models where migration rates depend on geographic distance. We previously found in a simple divergence model that $F_{ST} = T/(T + 2)$. If the expected divergence time between two populations is a linear function of the divergence time T, we then again find that there should be a linear relationship between geographic distance and $F_{ST}/(1 - F_{ST})$. So the same pattern of isolation by distance can be explained by both divergence models and migration models.

In humans, there is a clear pattern of isolation by distance (**Figure 4.8**). This pattern can be explained by both migration models and divergence models; it is often interpreted as being primarily a result of sequential colonization of areas out of Africa by modern human populations during the past 100,000 years. The degree to which human population structure can best be described by divergence models, models of ongoing gene-flow, or a mixture of the two has been a subject of intense debate among population geneticists, but sequential colonization is almost certainly an important part of the explanation.

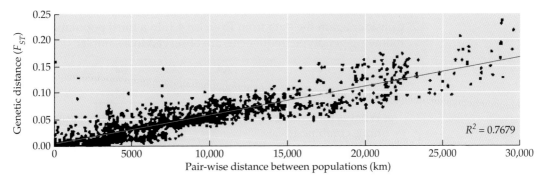

Figure 4.8 The relationship between F_{ST} and geographical distance between different pairs of human populations for more than fifty globally distributed populations. (After Handley et al., 2007.)

References

González-Suárez M., Flatz R., Aurioles-Gamboa D., et al., 2009. Isolation by distance among California sea lion populations in Mexico: Redefining management stocks. *Molecular Ecology* 18: 1088–1099.

*Handley L. J., Manica A., Goudet J., Balloux F., 2007. Going the distance: Human population genetics in a clinal world. *Trends in Genetics* 23: 432–439.

*Holsinger K. E. and Weir B. S., 2009. Genetics in geographically structured populations: Defining, estimating, and interpreting FST. *Nature Reviews Genetics* 10: 639–650.

*Myles S., Davison D., Barrett J., et al., 2008. Worldwide population differentiation at disease-associated SNPs. *BMC Medical Genomics*. doi: 10.1186/1755-8794–1–22.

*Slatkin M., 1991. Inbreeding coefficients and coalescence time. *Genetics Research* 58: 167–175.

*Valsecchi E., Palsbøll P. J., Hale P. T., et al., 1997. Microsatellite genetic distances between oceanic populations of the humpback whale (*Megaptera novaeangliae*). *Molecular Biology and Evolution* 14: 355–362.

*Recommended reading

EXERCISES

4.1 In two populations, the following genotype frequencies are observed in a sample:

	AA	Aa	aa
Population 1	20	20	20
Population 2	15	15	30

Calculate F_{ST} based on these samples.

4.2 Consider again the data from Exercise 4.1. In a third population, the following genotype frequencies are observed in a sample:

	AA	Aa	aa
Population 3	20	25	15

a. What is the average frequency of allele A in the three populations combined?

b. Calculate H_S, H_T, and F_{ST}.

4.3 A C/T polymorphism is segregating in two populations (populations 1 and 2) of spiny lizards. The two populations have been separated by a river restricting gene flow, but suddenly the river dries out, allowing gene-flow between the two populations at a rate of 0.1 per individual per generation. Before the river dried out, the frequency of the C allele was 10% in the first population and 90% in the second population.

> a. What is the expected frequency of C in population 1 one generation after the river dries out?
>
> b. What is it two generations after the river dries out?

4.4 Two populations share migrants at a rate of 0.0001 per gene copy per generation. Both populations have an effective population size of $2N = 10,000$.

> a. What is the expected coalescence time between two gene copies sampled from the same population?
>
> b. What is the expected coalescence time when sampling two gene copies, one from each population?
>
> c. Assuming an infinite sites model, what is the expected value of F_{ST}?

4.5 Two populations (population 1 and 2) diverged from each other 6000 generations ago, and since then there has been no migration between the two populations. The effective size of the population 1 is 10,000 and that of population 2 is 20,000. The effective size of the ancestral population from which the two populations diverged was 10,000. A researcher is studying a locus that mutates at a rate of 10^{-5} per generation and conforms to an infinite sites model.

> a. What is the expected number of nucleotide differences between two gene copies sampled from population 1?
>
> b. What is the expected number of nucleotide differences between a gene copy from population 1 and a gene copy from population 2?

4.6 Consider the data from Exercise 4.1 and assume a model with (equilibrium) migration between the two populations and equal population sizes of $2N = 10,000$. Provide an estimate of the rate of migration per individual per generation.

4.7 Consider again the data from Exercise 4.1. Now assume that the data were obtained from two populations of equal size, $2N = 10,000$, that diverged from a common ancestral population of the same size an unknown number of generations ago, with no gene-flow between them since the time of divergence. Provide an estimate of the number of generations since the two populations diverged from each other.

4.8 Using the result given in the text that $E_S[t] = d$ and $E_D[t] = \dfrac{1}{2M} + d$, show that $F_{ST} = \dfrac{(d-1)/d}{(d-1)/d + 2dM}$ under assumptions of an infinite sites model and an island model with d demes (populations), each of size $2N$, which exchange migrants at a rate m (= $M/2N$) per generation per deme.

4.9 Show that $E_S[t] = d$ and $E_D[t] = \dfrac{1}{2M} + d$ under assumptions of an infinite sites model and an island model with d demes each of population size $2N$, which exchange migrants pairwise at a rate m (= $M/2N$) per generation.

5 *Inferring Population History and Demography*

WHILE POPULATION GENETICS was a very theoretical discipline originally, the modern abundance of population genetic data has forced the field to become more data oriented. Population genetic data are now commonly used for estimating population sizes, charting the history of divergence and migration among populations, and for a large variety of other applications. In this chapter, we will introduce some of the commonly used methods for analyzing population genetic data. This chapter does not provide a detailed guide for how to analyze population genetic data; we will not make any references to particular computer programs or particular published methods. Rather we attempt to present the rationale underlying many of the methods used in population genetic data analysis. Coalescence theory has been particularly useful in these applications, because of its focus on the properties of samples taken from a population.

Inferring Demography Using Summary Statistics

To learn about historical and demographic processes, population geneticists build formal mathematical models, like the models presented in the previous chapters, and then devise methods for testing these models and for inferring the parameters of the models from the data. We have encountered several different parameters of population genetic models that are of interest to us, including θ ($= 4N\mu$) and M ($= 2Nm$) in models of gene-flow, and T (the divergence time between populations in number of generations divided by $2N$). We could also be interested in estimating other parameters—for example, parameters relating to changes in populations size. To make inferences about these parameters, we need to use a **statistic**. A statistic

Figure 5.1 Giant panda (*Ailuropoda melanoleuca*)

is anything that can be calculated from the data. Examples of statistics are the average number of pairwise differences and the number of segregating sites. In Chapter 3 we showed that the average number of pairwise difference and the number of segregating sites both could be used for estimating θ. For migration models, we can use an estimate of F_{ST}, calculated using the observed heterozygosity within and between populations, to estimate parameters such as M and T. For example, we saw that in the migration model with two populations, $M = (1 - F_{ST})/(8F_{ST})$. We can use this expression to convert an estimate of F_{ST} directly into an estimate of M. Likewise, in the divergence model, $F_{ST} = T/(T + 2)$, so T can be estimated using $T = 2F_{ST}/(1 - F_{ST})$. Other methods for estimating M from T have been proposed under different definitions of F_{ST} and using different population genetic models.

As a practical example, consider the management of endangered species such as the giant panda (*Ailuropoda melanoleuca*) (**Figure 5.1**). In order to assess the amount of migration between populations, He and colleagues analyzed DNA from the feces of giant pandas from two reserves (the Wangland and Baoxing) in China, and estimated F_{ST} to be 0.26. Assuming a simple island model with two islands, this would translate into an estimate for M of $(1 - 0.26)/(8 \times 0.26) = 0.36$ migrants per generation between the two populations. This kind of information is very useful to policy makers charged with protecting the species: it tells them that the two populations are genetically separated from each other and that genetic variability will be lost if one of the populations is allowed to go extinct.

Coalescence Simulations and Confidence Intervals

When estimating a parameter, it is often desirable not only to present the estimate itself, but also to give a measure of how confident you are in the estimate. In the previous section we obtained an estimate of $M = 0.36$ for the two populations of the giant panda, but if the estimate is based on very little data (e.g., very few SNPs or very few individuals) we may not have great faith that we would get the same answer were we to analyze another set of SNPs from the same populations with a study of the same size. A measure of confidence is needed to quantify the uncertainty in the estimate. However, in population genetics it is often difficult to provide simple measures of statistical uncertainty. Researchers therefore typically simulate new data to get an understanding of the variability in the estimate. An advantage of coalescence theory is that it allows fast and efficient simulations because it focuses on the history of just the sample, not the entire population. A coalescence-based simulation algorithm can be adapted to accommodate complex demographic histories including population splitting, gene-flow between populations, and changes in population size. It can also incorporate various forms of recombination between loci (see Chapter 6). By simulating new data, researchers can determine how likely it is that they would obtain similar estimates if they had sampled another set of DNA sequences from the same population(s). Coalescence simulations have, therefore, become fundamental to population genetic data analysis, and multiple programs are available for simulating samples.

Box 5.1 describes an algorithm for simulating coalescence trees under the standard neutral model. To generate data, it is necessary to assume a specific mutational model. For DNA sequence data, an infinite sites model is often assumed. Under this model, mutations are distributed evenly on all lineages at a constant rate $\theta/2$. When first mutations have been distributed on the tree, the data can be inferred directly from the tree (see **Figure 5.2**).

As an example, consider Tajima's estimator of θ ($\hat{\theta}_T = \pi$). For the data in Figure 5.2, we find that the average number of pairwise differences is 2.6. We can now simulate coalescence trees (using the algorithm in Box

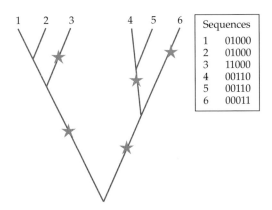

Sequences	
1	01000
2	01000
3	11000
4	00110
5	00110
6	00011

Figure 5.2 A coalescence tree with six leaf nodes representing six DNA sequences. Mutations in the coalescence tree are indicated in blue. Using an infinite sites model, the resulting DNA sequences (right) can be deduced by inspecting the tree. The DNA sequences are represented as binary sequences with ancestral alleles denoted by 0s and derived alleles by 1s. Because there are five mutations on the tree, the sequences contain five segregating sites (invariable sites are not shown). The order of the different sites is arbitrary.

BOX 5.1 Simulating Coalescence Trees

A coalescence tree can be described as a set of nodes with associated ages connected by edges (lineages). The ages of the nodes are defined to be zero for leaf nodes and equal to the respective coalescence times for the internal nodes in the tree. It can be simulated for a sample of n gene copies using an algorithm that starts from the leaf nodes and then simulates coalescence events recursively until the MRCA has been found:

Initialization:

- Set $k = n$, $T = 0$, and let $V = \{V_1, V_2, \dots, V_n\}$ be the set of leaf nodes with ages $A_1 = A_2 = \dots = A_n = 0$.

Recursion:

- Draw t from an exponential distribution with mean $2/[k(k-1)]$.
- Draw two nodes, V_a and V_b, uniformly from V such that $a \neq b$.
- Set $k = k - 1$ and $T = T + t$.
- Connect both V_a and V_b to a new node V_{2n-k} and set $A_{2n-k} = T$.
- Remove V_a and V_b from V and add V_{2n-k}.

Termination:

- Stop when $k = 1$.

In this algorithm, k represents the number of lineages currently in the tree and V is the set of nodes connected to these lineages. The recursion step is a loop that is repeated until the termination condition is satisfied, that is, until $k = 1$ and the MRCA has been found.

5.1), and distribute new mutations randomly on the lineages of these trees at rate of $2.6/2 = 1.3$ per coalescence time unit. The results of 100,000 such simulations are shown in **Figure 5.3**.

Notice that the estimates vary a great deal due to both the randomness introduced by the coalescence process and the randomness introduced by the mutation process. Among the simulated values, 2.5% are 0 and 97.5% are 7.27 or larger, so 95% fall in the interval [0, 7.27]. Sometimes such an interval, obtained by simulation, is used as an approximate **confidence interval**. In general, a 95% confidence interval is an interval that contains the true value of the parameter with 95% probability. How to form valid confidence intervals for many of the common estimators in population genetics is an area of active research.

Coalescence simulations can be used much more broadly than this to fit a model to the data. Using simulations, we can test whether the simulated data tend to fit the observed data under a particular model. When comparing two models, we would have more faith in a model that produces simulated data that look similar to the observed data than in a model that does not produce data that look anything like the real data. This basic concept is

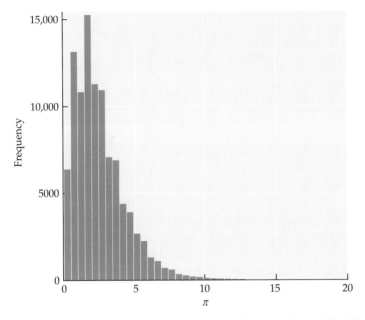

Figure 5.3 A histogram showing the distribution of values of π in 100,000 co-alescence simulations under the standard coalescence model with infinite sites mutation and $\theta = 2.6$.

used very often in population genetics, informally or more formally, and has made coalescence simulation one of the most important computational tools in population genetics.

Estimating Evolutionary Trees

We have discussed how to make inferences regarding population genetics using simple statistics such as S, π, and estimates of F_{ST}. But clearly, the data contain much more information than what is captured by such simple statistics. There is a long and well-justified tradition in evolutionary biology (phylogenetics) of focusing on evolutionary trees. We may want to do the same in population genetics, as the relationship between individuals with regard to any given locus in the genome is also represented by a tree: the coalescence tree. Phylogeneticists use several methods for estimating trees. The tree methods most commonly used fall into three groups: maximum parsimony methods, distance-based methods, and likelihood and Bayesian methods. We will not discuss these methods in detail, but will give a very brief overview of each group.

In the **maximum parsimony method**, the tree that requires the smallest number of mutations in order to explain the DNA sequence data is chosen. Consider, for example, the data in Figure 5.2. The data and tree in the figure are compatible with the infinite sites model and can be explained with

Figure 5.4 A tree and a set of binary sequences, which together are not compatible with the infinite sites model. SNPs that can be mapped on the tree with only one mutation are shown in red. The second and fourth SNP, in yellow and blue, respectively, require minimally two changes—for example, as indicated by the yellow and blue mutations on the tree to the left. So at least seven mutations are needed to explain the sequence data if this tree is the true tree.

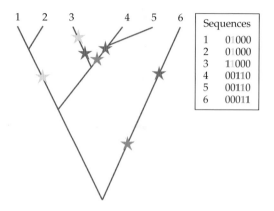

Sequences	
1	01000
2	01000
3	11000
4	00110
5	00110
6	00011

only five mutations. However, if we change to another tree, for example, a tree in which sequences 4 and 5 are grouped with sequence 3, then the tree and the data are no longer compatible with the infinite sites model (**Figure 5.4**), and 7 mutations are required to explain the data. According to the phylogenetic principle of maximum parsimony, we should prefer the tree in Figure 5.2 to the tree in Figure 5.4. To find the maximum parsimony tree(s), we always need to examine all possible trees and choose the tree(s) that require the fewest mutations.

Distance-based methods proceed by first estimating the genetic distance between all pairs of sequences. There are many different ways of doing this. We have encountered one such way already: estimating the number of pairwise differences. If the infinite sites model is a reasonable assumption, such a method will work well. But when comparing different species, the infinite sites model typically works very poorly, because there is a large chance that more than one mutation has hit each site. For this reason, statistical methods have been developed to estimate how many mutations have actually occurred between a pair of sequences (not just the number of observable pairwise differences). Estimating the number of mutations from the number of pairwise differences is sometimes called **correction for multiple hits**, because it takes into account the possibility that multiple mutations have hit the same nucleotide site. After estimating a distance matrix, a tree is then estimated which fits the distances as well as possible according to an algorithmic criterion. A computational advantage of distance-based methods is that they do not need to search all possible trees to find the best one. An example of a distance-based method is the UPGMA (Unweighted Pair Group Method using Arithmetic mean) method. The algorithm is briefly explained in **Box 5.2**. The UPGMA algorithm assumes a molecular clock—that is, it assumes that the mutation rate per year is the same in all lineages of the tree. Under this assumption, the distance from the root of the tree (the MRCA) is the same to all leaf nodes in the tree. When the molecular clock assumption is not met, because mutation rates differ between different lineages, then the UPGMA method does not tend to produce correct trees. In such cases, other algorithms, such as

BOX 5.2 The UPGMA Method for Estimating Trees

The UPGMA algorithm by Sokal and Michener (1958) for estimating trees from distance matrices assumes that distances between sequences have been calculated (a distance could, for example, be the number of nucleotide differences between the sequences). The algorithm proceeds very similarly to the algorithm in Box 5.1 for simulating coalescence trees. However, instead of choosing random nodes to coalesce, UPGMA chooses the nodes with the shortest distances between them. Also, the ages of the nodes in the tree are determined by the distances, and not by random simulation. The algorithm proceeds as follows:

Initialization:
- Set $k = n$, and let $V = \{V_1, V_2, \ldots, V_n\}$ be the set of leaf nodes with ages $A_1 = A_2 = \ldots = A_n = 0$. Let the distance between nodes i and j be $d_{ij} = d_{ji}$, $i, j = 1, 2, \ldots, n$.

Recursion:
- Identify the pair of nodes (a, b) in V with the smallest value of d_{ab}, $a \neq b$.
- Set $k = k - 1$.
- Connect both V_a and V_b to a new node V_{2n-k} and set $A_{2n-k} = d_{ab}/2$.
- Remove V_a and V_b from V and add V_{2n-k}.
- Define the distance from node $2n - k$ to any other node V_i in V, $i \neq 2n - k$, as the average distance between all descendent nodes of node $2n - k$ and all descendent nodes of node i.

Termination:
- Stop when $k = 1$.

the **neighbor-joining algorithm**, will work better. The neighbor-joining algorithm does not assume a molecular clock, but can take variation in the mutation rate among lineages into account. For most population genetic data, the rate of mutation is most likely very similar on the different lineages of the tree, at least if the particular loci analyzed have not been subject to natural selection.

The **maximum likelihood** and **Bayesian methods** are based on a likelihood function—the probability of the data given the parameters of a model, i.e., $\Pr(X | \Theta)$, where X represents the data (e.g., a set of DNA sequences) and Θ (theta) symbolizes the parameter we wish to estimate (in this case, the tree). The vertical bar is read as "given," and indicates that we wish to calculate the probability of the data given a specific set of parameters (see Appendix A). The likelihood function can be calculated for a specific model of molecular evolution using standard computational methods. The **maximum likelihood principle** then tells us that we should prefer the tree which gives the highest value of the likelihood function, i.e., the value of θ that maximizes $\Pr(X | \Theta)$. Notice its similarity to the maximum parsimony method: in both, we search among all trees to find the tree that maximizes or minimizes some criterion. Maximum parsimony chooses the tree that requires the fewest number of mutations, and maximum likelihood chooses the tree that maximizes the likelihood of observing the data. Generally, if the model used to describe the process of mutation is correct, the maximum likelihood method is likely to perform well and do better than other methods. In fact, it is a general principle in statistical inference that maximum likelihood is the optimal method if the data set is large and the model is correct. However, if the model is flawed, the method may not perform better than, say, the maximum parsimony method.

In Bayesian phylogenetic inference, the objective is to estimate the probability that a particular tree is the correct tree. This is also done using the likelihood function; but in addition, a prior distribution, $Pr(\Theta)$, is assumed. The prior distribution is a probability distribution that quantifies the researcher's belief in different trees before analyzing the data. For example, if the researcher believes the two trees in Figures 5.2 and 5.4 are equally likely, they should be given the same prior probability. A posterior probability is then calculated by multiplying the prior probability and the likelihood function, and dividing by a constant. The posterior probability gives $Pr(\Theta \mid X)$, i.e., the probability (or probability density) of the parameter given the data. In phylogenetic inference, it is the probability of the tree given the information obtained from the data. Calculating the posterior probability is not easy, but is done using a simulation technique called **Markov Chain Monte Carlo (MCMC)**. In Bayesian phylogenetics, the best tree is usually the one with the highest posterior probability, but there are also other methods for choosing the best tree using posterior probabilities. Like maximum likelihood methods, Bayesian methods will tend to perform well when the assumptions regarding the underlying models are met. Maximum likelihood estimation and Bayesian estimation are discussed in more detail in Appendix D.

Which method to use in phylogenetic inference has been a contentious issue over the past four decades. Methods based on optimization or simulation (maximum parsimony and likelihood-based methods) can be very slow, because the number of possible trees is typically very large, and it can be very difficult computationally to find the optimal tree. Likelihood-based methods (maximum likelihood and Bayesian methods) are particularly slow, because calculation of the likelihood function is in itself very slow. The choice of method for estimating trees is, therefore, often a pragmatic choice weighing what is theoretically optimal against what is computationally feasible. This is especially true for large data sets.

Gene Trees Versus Species Trees

The phylogentic methods discussed in the previous section have been developed primarily for estimating phylogenies, i.e., for elucidating the patterns of species evolution. However, they are now also commonly used to analyze population genetic data. Before venturing further into the use of estimated trees in population genetics, it might be appropriate to discuss the relationship between species phylogenies and gene trees estimated from DNA sequence data or other genetic data. **Figure 5.5** graphically presents a model of diverging populations. While speciation is a complex process

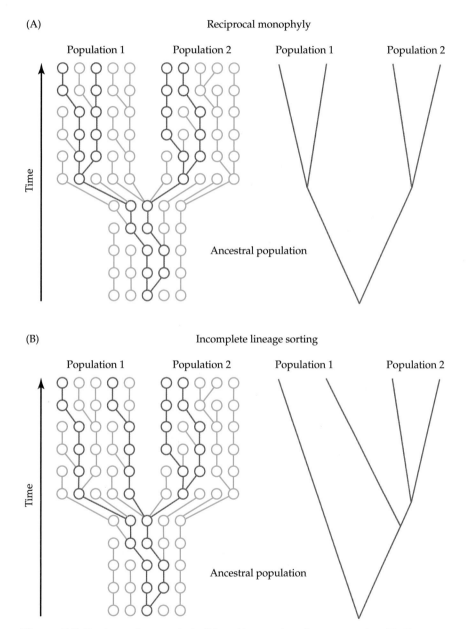

Figure 5.5 Reciprocal monophyly (A) and incomplete lineage sorting (B). Two gene copies have been sampled from each population, and the ancestry of the entire sample is traced back to the MRCA. The lineages that are part of the ancestry of the sample are marked in red.

that may not involve a discrete splitting event like that shown in Figure 5.5, a model of diverging populations may be a good first approximation for species evolution. We may think of the two populations in Figure 5.5 as representing two different species, say humans and chimpanzees. In this case, T represents the divergence time between species, while t is the coalescence time between two gene copies (e.g., DNA sequences), one sampled from each species. When estimating gene trees from DNA sequences, we estimate coalescence times, not divergence times. Since $t > T$, we tend to overestimate the species divergence time when estimating trees from DNA sequences. How important this problem is depends on the effective population size of the ancestral population and the divergence time (T). If T is large and the ancestral population size is very small, then t and T will be approximately equal. But if T is small and the ancestral population size is large, there can be a substantial difference between the estimated coalescence time and the species divergence time. For closely related species, the estimated time to the MRCA should not be confused with an estimate of the species divergence time. The time to the MRCA will depend on the amount of genetic variation in the ancestral species.

The problem of ancestral variation affects not only the estimates of the divergence time, but also the structure of the tree itself (the topology). A set of leaf nodes in a tree is said to form a monophyletic group if they share an MRCA with each other that is not shared with any other leaf nodes. When all the individuals within each species share an MRCA with each other that is not shared with individuals outside the species, i.e., the individuals within each species form a monophyletic group, the we say that there is **reciprocal monophyly** (Figure 5.5A). In this case, the species tree and the gene tree are concordant—they have the same structure no matter which individuals we sample.

In Figure 5.5A the two lineages from each species coalesce before (looking back in time) the divergence of the two species. However, if the divergence time is short and the population sizes are large, this may not necessarily happen. With some probability, the individuals in the sample from each population may not have an MRCA by the time of divergence, i.e., more than one lineage may survive. If this happens, there is some chance that the subsequent coalescence process in the ancestral population will generate trees that do not show reciprocal monophyly, that is, individual(s) from one species share an MRCA with individuals from the other species, not shared with the other members of their own species. Population geneticists call this **incomplete lineage sorting** (Figure 5.5B).

If more than two species have been sampled, the picture gets more complicated, but if the internal lineages in the species tree are short relative to the population size, the coalescence tree may no longer match the species tree. This is illustrated in **Figure 5.6**. In the case of the red lineages, the lineages from species 2 and species 3 coalesce in their common ancestor, species A_{23}. This ensures that the coalescence tree will match the species

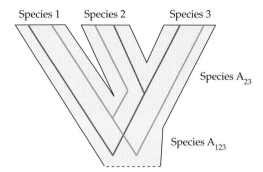

Species 1   Species 2   Species 3

Species A$_{23}$

Species A$_{123}$

Figure 5.6 The coalescence tree may (red) or may not (blue) match the structure of the species tree (black). The ancestral species of species 1 and 2 are labeled A$_{23}$, and the ancestral species for all three species is labeled A$_{123}$.

tree. However, in the case of the blue lineages, an MRCA for the lineages from species 1 and 2 is not found in species A$_{23}$; the two lineages do not coalesce. Consequently, there are three ancestral lineages in species A$_{123}$ (the ancestral species common to all three species). This allows the lineage from species 2 to coalesce with the lineage from species 1 before the lineage representing their shared ancestors coalesces with the lineage from species 3. In the case of the blue lineages, the coalescence tree is not congruent with the species tree.

Under the standard assumptions used in Chapter 3 to derive the coalescence process, we can relatively easily calculate the probability of incongruent trees for three species when sampling one gene copy (e.g., DNA sequence) from each species. If no coalescence event has happened in the ancestral species (as for A$_{23}$ in Figure 5.6), three different coalescence events could occur in species A$_{123}$—between lineages 1 and 2, 1 and 3, or 2 and 3. Each of these events may happen with equal probability, and two of the three possible coalescence events will lead to an incongruent tree, so the chance of an incongruent tree structure is thus ⅔.

Recall from Chapter 3 that the probability that two lineages do not coalesce during t time units, where t can be any non-negative value, is $e^{-\tau}$ under the standard assumptions of the coalescence process (Equation 3.4). So if species A$_{23}$ has a constant population size of $2N$ gene copies (N diploid individuals), and a branch length of $\tau 2N$ generations, i.e., τ is the branch length of population A$_{23}$ scaled by the population size, the chance of no coalescence event between lineages 2 and 3 while this species persists is e^{-t} (see Chapter 3). The total probability is then

$$\Pr(\textit{incongruence between gene tree and species tree}) = 2e^{-\tau}/3 \qquad (5.1)$$

When τ is sufficiently small (e.g., < 10), the topology obtained may depend on which individuals have been sampled to represent the species and which genes have been chosen. A point that we will return to later in this chapter is that because of recombination (see Chapter 6), the coalescence tree will be different for different loci in the genome. So some genes in the genome may show congruence between gene tree and species while others do not.

The tree relating humans and the great apes used to be unresolved, with various evidence suggesting that humans and chimpanzees, or gorillas and chimpanzees, or gorillas and humans form a monophyletic group. Thanks to phylogenies based on DNA sequence data, it is now widely accepted that it is humans and chimpanzees who are each other's closest relatives. One of the reasons there has been so much debate about this phylogeny is that the ancestral species lineage leading from the ancestor of all three species to the ancestor of humans and chimpanzees is very short. It has been estimated that due to incomplete lineage sorting, only $2/3$ of the gene trees in the nuclear genome follow the species tree in this case. In approximately $1/6$ of the genome, we are more closely related to gorillas than to chimpanzees, and in $1/6$ of the genome, gorillas and chimpanzees are more closely related to each other than to us.

It should be noted that there are many reasons other than incomplete lineage sorting for a discrepancy between species trees and estimated DNA sequence trees. The most obvious is estimation uncertainty. If the data used to estimate the coalescence tree are limited, the estimates of the tree are not likely to be very accurate. So an apparent lack of concordance between the coalescence trees may simply be an estimation artifact. Also, we have been appealing to a rather essentialistic view of species in this section as discrete units cleanly separated from each other at distinct points in time. Real species may occasionally share some limited gene-flow even long time after the first time of separation. Phylogeneticists sometimes call this **horizontal gene transfer**. When some limited gene-flow remains between species, this may also cause discrepancies between estimated coalescence trees (gene-trees) and species trees. Horizontal gene transfer may affect the human/chimpanzee/gorilla tree; it has been hypothesized that substantial gene-flow remained after the initial divergence between gorillas and the ancestor of humans and chimpanzees.

Interpreting Estimated Trees from Population Genetic Data

Estimation of a tree is just the first step of tree-based inference on population genetics. The second step is to interpret the tree in terms of population genetics. As we shall see, that is not always as easy as one might think.

Consider, for example, the mtDNA tree in **Figure 5.7**. Several features of this tree are interesting to anthropologists. Most importantly, the root of the tree (determined by comparing the human mtDNA sequences to the mtDNA of an outgroup such as the chimpanzee) falls within African variation. The tree is compatible with a model of human history in which humans originated in Africa and then moved out of Africa and colonized the rest of the world. The high degree of variability in Africa compared with other parts of the world is consistent with a scenario in which the non-African population(s) went through a bottleneck (a strong short-term reduction in population size) during the out-of-Africa migration event.

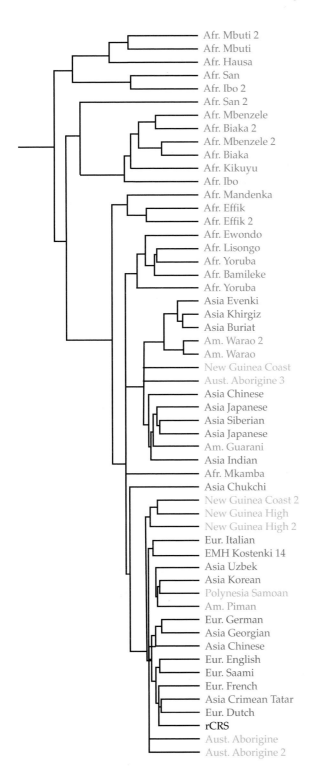

Figure 5.7 Human mtDNA tree. The sampling includes Africans (blue); Asians (red); Native Americans (green); Europeans (purple); and Australian Aborigines, Polynesians and Melanesians (orange). rCRS is the human reference mtDNA sequence. EMH Kostenki 14 is the mtDNA from the 30,000 year old remains of a Siberian individual. Notice that the root of the tree is placed within Africans. Also notice that there generally is not reciprocal monophyly between different continental groups. (After Krause et al., 2010.)

Different historical and demographic models make different predictions about the underlying coalescence trees. Sometimes the predictions are very clear; other times the relationship between coalescence trees and demographic models is more opaque. For example, a tree such as that in **Figure 5.8A**, showing clear reciprocal monophyly, is most likely to occur when the divergence time between populations is large and there has been very little gene-flow, if any, since they diverged. In contrast, if the two populations diverged from each other very recently, and still share a very high level of gene-flow and **panmixia**—random mating between individuals from population 1 and 2—we would expect to see a tree such as that in **Figure 5.8B**. If the divergence time is long, and gene-flow has been limited, but there has been some very recent gene-flow, a tree such as that in **Figure 5.8C** would be expected; there will be some recent coalescence events between lineages from population 1 and population 2. However, if gene-flow has been ongoing at low levels for a long of time between populations on an island, coalescence trees as in **Figures 5.8D, E,** and **F** might be expected, in which one or a few lineages cross between populations. In Figures 5.8E and F, we would expect the effective population size of population 1 to be larger, as it has an older MRCA. But Figures 5.8D, E, and F might also be compatible with models of divergence between populations without any subsequent gene-flow after the time of divergence. Figure 5.8D would be entirely compatible with a model of recent divergence, and incomplete ancestral lineage sorting. Figure 5.8E looks exactly like the expected tree sampled from two populations in which the second population is derived from the first through a bottleneck event, or in which the second population has a much smaller effective population size than the first population. Figure 5.8F is also compatible with a model of divergence between populations 1 and 2 without subsequent gene-flow. However, in this case, all ancestral variation has not been eliminated in population 2 by the bottleneck or generally low effective population size, possibly because the divergence event happened so recently that there is still some residual incomplete lineage sorting.

From Figure 5.8 it should be clear that the demographic history influences the shape of sampled trees. Therefore, inferences can be made regarding the demographic history of populations examining coalescence trees. But it should perhaps also be clear that there is not a simple one-to-one relationship between models and trees. For example, all the trees in Figure 5.8 could be generated with a model of divergence without gene-flow and a model with ongoing gene-flow, although they may not all be equally likely under both models.

The reader may notice that the general structure of the tree in Figure 5.8F is similar to that of the human gene tree (Figure 5.7), if we let population 1 represent Africans and population 2 represent non-Africans. Much emphasis has been made on the fact that the root of the human tree (the MRCA of all humans) falls within African variation, i.e., that non-Africans have an MRCA that is more recent than the MRCA of Africans. It has been interpreted as

(A) Old divergence, little gene flow

Population 1   Population 2

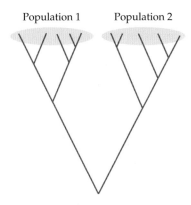

(B) Strong gene-flow, panmixia,
very recent divergence

Population 1   Population 2

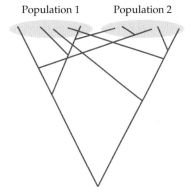

(C) Old divergence, recent gene-flow

Population 1   Population 2

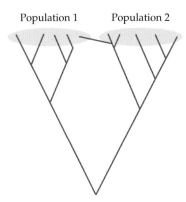

(D) Ongoing gene-flow, old divergence
or recent divergence

Population 1   Population 2

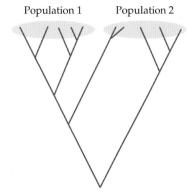

(E) Old divergence or ongoing gene-flow,
low N_e in population 2

Population 1   Population 2

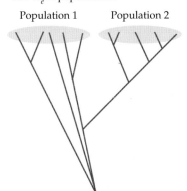

(F) Recent divergence or ongoing gene-flow,
low N_e in population 2

Population 1   Population 2

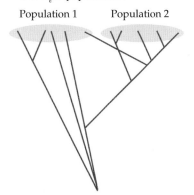

Figure 5.8 Coalescence trees produced by different demographic and historical processes.

strong evidence that humans originated in Africa and later migrated out of Africa. This perception of human evolution is also in agreement with the dominant view of archaeologists and paleontologists. However, a tree with a root in Africa could potentially be consistent with other population genetic models. For example, a model with gene-flow with a larger effective population size in Africa than in the populations outside Africa would also be likely to produce such a tree. Such a model is compatible with the so-called **multiregional hypothesis** of human evolution, which assumes that modern humans evolved simultaneously in many regions of the world, but with some gene-flow between the different regions.

This discussion of gene trees in populations should illustrate that direct inference of demographic history from a single estimated gene tree is not easily done. There might be several models of the history of the population than can explain the same tree. Furthermore, a particular demographic history can produce many different trees, due to the stochastic nature of the coalescence process. Basing demographic inferences on a single recombining unit of DNA, such as mtDNA and Y chromosome DNA, has therefore great potential to be misleading if not interpreted in the context of coalescence theory. In order to quantify how likely a tree is under a given population genetic model, we need to be able to infer, somehow, which demographic scenario produced the tree. Statistical methods for doing this will be the topic of the remainder of this chapter. Much of this material is more advanced than other topics in this book and can be skipped by readers not interested in these issues.

Likelihood and the Felsenstein Equation

We have considered two types of approaches to making inferences in population genetics from DNA data. In the first approach, we use simple summary statistics such as F_{ST} and π to estimate parameters of population genetic models. In the second type of approach, we attempt to estimate the coalescence tree and base our inferences on this estimated tree. The first approach has the drawback that we lose information about the population by reducing the data to simple summary statistics. For example, while we can estimate either migration rate (M) or divergence (T) based on such approaches, they cannot help us to determine if a model that assumes divergence between populations with little or no gene-flow after divergence or a model of ongoing gene-flow in an island best describes the data. By estimating a tree, we can distinguish between such models. However, the tree-based approaches also suffer from serious drawbacks. In particular, as illustrated in the previous section, it is not always clear exactly which models may be compatible or incompatible with a particular tree. In addition, there is statistical uncertainty in the estimation of the tree. Population genetic inferences based on an estimated tree are only as good as the tree is.

In the mid-1990s, population geneticists realized that there was a need for better methods that would allow for clear interpretations, as the summary statistics do, but which also could take advantage of all the information in the data—methods that make use of the coalescence trees, and are based on a solid statistical footing. Statistical theory tells us that if we want to use all information in the data, we should base our inferences on the likelihood function. The likelihood function was previously discussed in this chapter in the context of phylogenetic inference, and it is discussed in more detail in Appendix D. In brief, the likelihood function is the probability of the data given the parameter values, $\Pr(X \,|\, \Theta)$, where X represents the data (e.g., a set of DNA sequences), Θ is a vector of parameters (e.g., migration rates and effective population sizes), and the vertical bar is read as "given" or "conditional on." The likelihood function provides information about how likely the observed data are for particular values of the parameters. The most common method for estimating parameters using the likelihood function is to use maximum likelihood, that is, to choose the values of the parameters that give the highest probability of observing the data. Clearly, if one set of parameters—for example very low migration rates—is very unlikely to have produced data similar to the observed data, but another set of parameters—for example, high migration rates—is very likely to produce data similar to the observed data, we would tend to trust a model with high migration rates more than a model with low migration rates. Population geneticists have, therefore, vigorously pursued methods for calculating the likelihood function in population genetic models.

As a simple example, consider the expression given in Equation 3.17 for the probability of obtaining S differences between two sequences under the standard coalescence model with infinite sites mutation. We see that this is a function of θ, and therefore is a likelihood function for θ. The function is shown in **Figure 5.9** for the case of $S = 6$.

If we do the calculus to find the value of θ that maximizes the likelihood function, we find it to be S (See Appendix D). So $\hat{\theta}_{ML} = S$ is the maximum likelihood estimator of θ. For the example in Figure 5.9, the maximum likelihood estimate of θ is $\hat{\theta}_{ML} = 6$. This is, in fact, the very same estimator of θ as the two previously encountered estimators of $\hat{\theta}_T$ and $\hat{\theta}_W$ for $n = 2$. If n is larger than two, then the three estimators will be different from each other.

Unfortunately, most of the time, it is not easy to calculate the likelihood function. Complicated simulation approaches are most often required. Most of these simulation approaches can be thought of as applications of the **Felsenstein equation**, named after the famous population geneticist and phylogeneticist, J. Felsenstein:

$$\Pr(X \,|\, \Theta) = \int_G \Pr(X \,|\, G) p(G \,|\, \Theta) dG \tag{5.2}$$

This equation may look complicated at first. We see that it involves an integral over all possible values of G. G represents the coalescence tree. The integral is evaluated by examining all possible coalescence trees, and for each

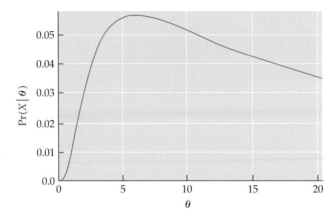

Figure 5.9 The likelihood function for θ under the standard coalescence model with infinite sites mutation when $n = 2$ and the two sequences differ by six nucleotide sites.

of the trees, integrating over all possible branch lengths of the tree, while multiplying the functions $p(G\,|\,\Theta)$ and $\Pr(X\,|\,G)$ with each other. $p(G\,|\,\Theta)$ is the distribution (density) of coalescence trees given the parameters (such as population sizes, migration rates, divergence times, etc.). This function can be calculated using coalescence theory. We know the distribution of coalescence times, and from this distribution the entire distribution of co-alescence tree can be derived. $\Pr(X\,|\,G)$ is the probability of the data given a particular tree. In phylogentics there is a well-developed theory for how to calculate $\Pr(X\,|\,G)$, which is used extensively in maximum-likelihood estimation of phylogenetic trees, so this part of the function is also easy to calculate using standard methods.

The evaluation of the likelihood function involves a consideration of all trees. Instead of just concentrating on one possible estimated tree, the likelihood function considers all possible trees and weights them by their relative likelihood. This approach circumvents the problem of estimation uncertainty in the tree and provides a rigorous method for relating the data to population genetic models and hypotheses. The likelihood function can also be used to test different hypotheses. We will not go into detail with the statistical methods used for testing hypotheses using likelihoods. However, it is clear that we should believe more in models with a high likelihood than in models with a very low likelihood. This basic principle can be extended to provide formal statistical tests of specific models and to discriminate between different models using DNA data.

MCMC and Bayesian Methods

Unfortunately even though both $\Pr(X\,|\,G)$ and $p(G\,|\,\Theta)$ are usually easy to calculate, the integral itself is not, because it involves evaluating all pos-

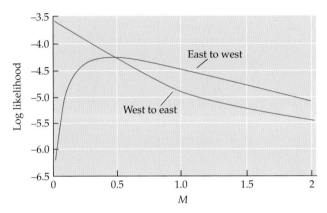

Figure 5.10 Likelihood surfaces for the migration rate parameter M (= 2Nm) for two populations of sticklebacks from the Western and Eastern Pacific Ocean. The logarithm of the likelihood is shown instead of the likelihood itself. The rates of migration from east to west and from west to east are shown. Notice that the likelihood surface for migration from west to east is a strictly decreasing function, implying that there is no migration from west to east (the maximum likelihood estimate of the migration rate equals zero). However, the likelihood surface for the migration rate from east to west has a maximum approximately at M = 0.5, suggesting gene-flow at a rate of one migrant every second generation from the Eastern Pacific to the Western Pacific. (From Nielsen and Wakeley, (2000) based on data from Orti et al., 1994.)

sible coalescence trees for all possible branch lengths; in practice, an impossible task under most conditions. Most of the time, various simulation approaches are used for this. The basic idea in the simulation approaches is to just evaluate some of the possible trees, but to do it in such a way that these trees are representative for all possible trees. This can lead to very accurate evaluations of Equation 5.2. The most commonly used technique for doing this is Markov Chain Monte Carlo (MCMC), which is a standard technique from computational statistics used to approximate distributions by simulation. A full exploration of MCMC methods is beyond the scope of this book. However, an example of likelihood functions calculated by MCMC is shown in **Figure 5.10**. There are several alternative simulation methods for MCMC, including the simulation method of R. C. Griffiths and S. Tavaré (see Recommended Readings).

In some special cases, the likelihood function can be calculated without using simulation. For the infinite sites model when n is small or when there are only very few segregating sites, various computational methods can be used to calculate the likelihood function directly. Under the infinite alleles model (see Chapter 3) the likelihood function for θ can be calculated directly using the so-called **Ewens sampling formula**, named after the famous population geneticist W. Ewens.

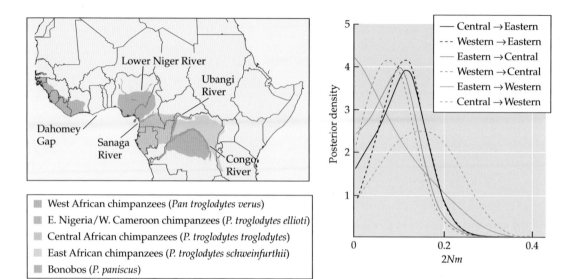

Figure 5.11 A map of the distribution of different chimpanzee subspecies and the posterior distribution of the migration rates between Eastern, Central, and Western chimpanzees (*Pan troglodytes*), estimated by Hey (2010) using an MCMC method. The level of migration is quite low for all populations (2*Nm* < 0.4). The most probable values of 2*Nm* are around 0.1 for most population pairs, except for migration from Eastern to Western chimpanzees, for which the most probable value is zero.

An alternative to maximum-likelihood estimation is to use Bayesian methods for inference. Bayesian methods were discussed in the section regarding phylogenies, and are also discussed in more detail in Appendix D. In brief, the objective is to estimate a **posterior distribution** of the parameter. The posterior distribution provides the probability distribution of the parameter, given data $f(\Theta \mid X)$ for continuous parameters and $\Pr(\Theta \mid X)$ for discrete parameters using the notation of Appendix A. For example, if we wish to estimate θ, the posterior distribution would summarize the belief we have in θ taking on any particular value. The value with highest posterior probability would be the value we would have the strongest belief in. In order to calculate the posterior distribution, we also need to make assumptions about the **prior distribution**. The prior distribution summarizes our information regarding the parameter before we have observed any data. Typically, we have very little prior information about parameters such as θ, T, or M, and would therefore assume a uniform distribution, which puts equal weight on all possible values of the parameter. The previously discussed MCMC methods are usually constructed so that they directly estimate the posterior distribution. An example of posterior distributions of the migration parameter $M = 2Nm$, estimated using MCMC, is shown in **Figure 5.11**.

Because it often can be computationally difficult, even using simulation methods such as MCMC, to evaluate the likelihood function (or posterior probabilities), a number of approximation methods have been developed. These methods trade a reduction in statistical accuracy for faster computational time. The most popular method is called **Approximate Bayesian Computation (ABC)**. This method aims at calculating the posterior distribution of the parameter of interest, but it does so by using only some of the information in the data. For example, imagine we wish to estimate θ from DNA sequence data, assuming the standard coalescence model and infinite sites mutation, but we want to do so using both the information from the average number of pairwise differences (π) and the number of segregating sites (S). An ABC method would then proceed by simulating data for various values of θ chosen randomly from the prior distribution. For each value of θ, there is a corresponding simulated data set for which π and S can be calculated. Simulated data sets are either accepted or rejected, depending on how different the simulated data set is from the observed data. In our example, we might accept a data set if the simulated values of π and S are sufficiently close to the observed values. The distribution of values of θ for the accepted data sets, can then be shown to approximate the posterior distribution of θ based on π and S.

The Effect of Recombination

So far we have assumed that there is a single coalescence tree describing the gene genealogy of the data. This will generally not be the case in the presence of recombination. Recombination is discussed in greater detail later in the book. Here we will mainly be interested in one consequence of recombination: when there is recombination between different loci, then the coalescence trees will differ among loci. In any genome, there will be thousands or even millions of different coalescence trees, each tree being specific to a particular segment of the genome. In many organisms, recombination events happen just as frequently as mutations. This implies that the theory and methods discussed so far are inapplicable to genomic segments of any significant length. Notable exceptions include mtDNA and Y chromosome DNA, which do not undergo recombination.

Even though there is not a shared coalescence tree for the entire genome, one can still estimate a tree. The tree then represents the average coalescence times between sequences ($2N$ in a standard coalescence model). If there is no population structure, we would expect all individuals—on average, when looking at many regions of the genome—to be equally closely related to each other. As a result, we would also expect the underlying tree to have the structure of a **star phylogeny (Figure 5.12)**, a tree in which all individuals are equally close to each other and all the internal lineages all are of zero length. Any internal lineage of a length greater than zero would indicate differences in average coalescence time between different individuals, and be indicative of some degree of population structure.

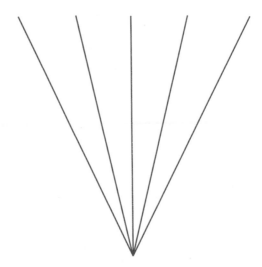

Figure 5.12 A star phylogeny: the expected average tree when many loci from a randomly mating population are analyzed simultaneously. Notice that such a tree is quite different from a standard coalescence tree.

Until recently, population genetic analyses focused primarily on one or a few loci, but now genome-wide data with thousands of Single Nucleotide Polymorphisms (SNPs) are being generated for many different species. Most of the tree-based methods used for estimating population genetic parameters, including the MCMC methods, are not readily applicable to this type of data. These methods rely on an explicit representation of a coalescence tree. But when each nucleotide site may have its own tree, and when the data include thousands of nucleotide sites, it becomes impractical to rely on methods that explicitly consider the coalescence tree. Fortunately, we can use a number of methods that do not assume all site have the same tree, and which can be applied to thousands of sites simultaneously. Many of these methods focus on the Site Frequency Spectrum (SFS; see Chapter 3). For example, consider the case of changes in population size. As discussed in Chapter 3, coalescences tree in growing populations tend to have relatively long external branches, and coalescence trees in populations with declining population sizes tend to have relatively short internal branches. Trees with longer external branches produce more singletons, and trees with shorter external branches produce fewer singletons. So changes in population size leave a specific pattern in the SFS that can be used to estimate how the population size has changed through time. For example, many human populations tend to have more singletons than expected under the standard coalescence model, because they have experienced population growth (**Figure 5.13**). By fitting the expected frequency spectrum under a particular model to the observed frequency spectrum, the amount of population growth can be quantified.

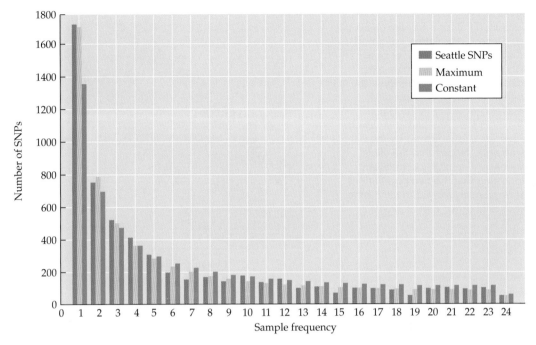

Figure 5.13 The site frequency spectrum (SFS) for a sample of African Americans for 5982 SNPs from Adams and Hudson (2004). The observed data are represented by the bars labeled "Seattle SNPs." The expected distribution under the standard coalescence model is labeled "Constant." The expected SFS under a model fitted to the data, which assumes an increase in population size $0.54 \times 2N$ generations ago, is labeled "Maximum."

When estimating parameters from more than two populations—for example, migration rates or population divergence times—the **joint frequency spectrum** can be used (**Figure 5.14**). The joint frequency spectrum summarizes the distribution of allele frequencies in two or more populations. Again, the expected frequency spectrum can be fitted for various models to the frequency spectrum observed in the data to estimate parameters of a demographic model including divergence times, migration rates, and effective population sizes.

Population Assignment, Clustering, and Admixture

In Chapter 1 we discussed match probabilities—the probability of random identity between two DNA profiles. Match probabilities can also be used to assign specific individuals to populations. Consider an example in which there are two populations, 1 and 2, with allele frequencies f_{A1} and f_{A2}, respectively, at locus A. Assume also that we have observed an individual with genotype AA. Under the assumption of HWE, the match probability would be $\Pr(genotype = AA \mid pop = 1) = f_{A1}^2$ if the individual belongs to population 1,

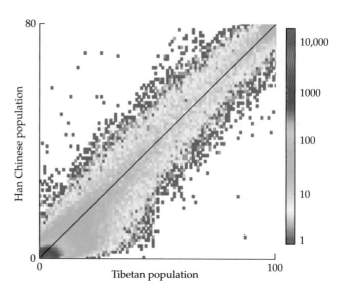

Figure 5.14 The joint site frequency spectrum (SFS) for a Tibetan and a Han Chinese population estimated for a genome-wide data set of all protein-coding genes. The x-axis indicates the allele frequency in the Tibetans and the y-axis indicates the allele frequency in the Hans. The colors show the particular allele frequencies, as indicated by the bar to the right. Notice that the allele frequencies are highly correlated, suggesting that the two populations are very closely related to each other genetically. (After Yi et al., 2010.)

and $\Pr(genotype = AA \mid pop = 2) = f_{A2}^2$ if the individual belongs to population 2. If f_{A1} and f_{A2} are different from each other, the match probabilities will differ depending on which population we assume the individual to originate from. For example, if $f_{A1} = 0.1$ and $f_{A2} = 0.9$, the match probabilities are 0.01 and 0.81, respectively. Clearly, it is more likely that the individual comes from population 2 than population 1. We can use **Bayes' law** (Appendix A) to calculate the probability that the individual comes from population 1 or population 2. To do so, we need to make an assumption abut the prior probability that the individual comes from population 1 or from population 2. In the absence of any other information, it makes sense to assume that the probability that the individual comes from population 1 equals the probability that the individual comes from population 2, i.e., $\Pr(pop = 1) = \Pr(pop = 2) = 1/2$. We then have

$$\Pr(genotype = AA \mid pop = 1) = f_{A1}^2, \Pr(genotype = AA \mid pop = 2) = f_{A2}^2$$

$$\Pr(pop = 1 \mid genotype = AA)$$

$$= \frac{\Pr(genotype = AA \mid pop = 1)\Pr(pop = 1)}{\Pr(genotype = AA \mid pop = 1)\Pr(pop = 1) + \Pr(genotype = AA \mid pop = 2)\Pr(pop = 2)}$$

$$= \frac{f_{A1}^2 \times 0.5}{f_{A1}^2 \times 0.5 + f_{A2}^2 \times 0.5} = \frac{f_{A1}^2}{f_{A1}^2 + f_{A2}^2} \tag{5.3}$$

If the allele frequencies are known in each population, we can use the genotype of a sampled individual to determine the probability that the individual belongs to a particular population. The evidence from multiple loci

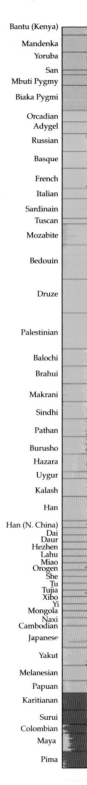

Figure 5.15 Admixture analysis of 1056 individuals from 52 populations for 377 microsatelite loci. Each individual is represented by a vertical line. The contributions to the genetic ancestry are shared by four populations, represented by the four colors—orange, blue, pink, and purple. The color of each line represents the proportion of the individuals' genetic ancestry that is due to that population. For example, all African individuals descend predominantly from the orange population. All individuals outside Africa have very little or no genetic ancestry from the orange populations, except for a few individuals from the Middle East. All European individuals descend almost exclusively from the blue population, and so on. (After Rosenberg et al., 2002.)

can be combined by multiplying the match probabilities calculated for each locus into a single combined match probability. This type of multiplication is valid if the loci are not closely linked together. In this way, even small differences in allele frequency can translate into strong evidence regarding the origin of an individual when the data from multiple loci are combined.

It is also possible to model the situation in which an individual has some genetic ancestry from multiple populations—that is, their ancestry is **admixed**. The match probability can be calculated by averaging over the contributions from the different populations. An example is shown in **Figure 5.15**. A fraction of each individual is assigned to a population. We see in Figure 5.15 that the categorization of humans reasonably closely follows the major continental groups. Even though most of the genetic variability in humans is not due to population subdivision between continental groups, it is still possible to assign individuals quite accurately to each group. In fact with genomic data, humans can be assigned to geographic regions with, at times, surprising accuracy.

There are a number of other different population genetic methods for analyzing genomic data. There are also various methods for defining genetic distances among individuals, and then depicting the genetic relationships of individuals using trees or other types of plots. The most commonly method is called **Principal Component Analysis (PCA)**. PCA is a commonly used statistical method for identifying features that are important in high-dimensional data. A full description of the method is beyond this book, but in genetics it is typically used as a method for clustering individuals and identifying those who belong to similar groups, much like the population assignment and admixture analyses. A number of principal components are estimated. Each component summarizes some features of the data that allow discrimination among individuals. Typically the first few components are considered, and are used to generate a graphical depiction of the relationship among individuals. According to the axes of variation defined by the PCA, individuals close to each other in the graphical depiction are more closely related than individuals distant from each other. In this way, the principal components are used to define genetic distances based on thousands of SNPs (**Figure 5.16**).

Figure 5.16 A PCA analysis of 3000 European individuals, using 500,000 SNPs for each individual. The color labeling and acronyms for countries are explained by the European map in the top right-hand corner. Notice that with a proper transformation, the distances between individuals, as summarized by the first and second principal components (PC1 and PC2), come remarkably close to mirroring the geographic distances between sampled individuals. (After Novembre et al., 2008.)

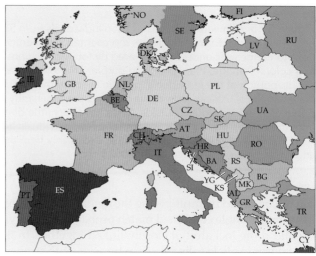

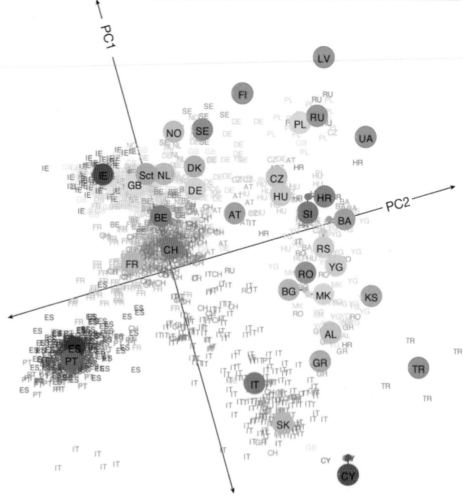

PCA analyses, population assignments, and admixture analyses are very powerful methods for clustering individuals and exploring the genetic relationships among individuals for large SNP data sets. But the results are not easily interpretable in terms of population genetics. These methods are agnostic regarding the processes causing differences among individuals. They do not provide estimates of divergence times or migration rates, and cannot be used directly to infer the demographic history of the populations analyzed. They find their strength when populations are not well defined and exploratory analyses are needed to define natural units for population genetic analyses.

References

*Adams A. M., Hudson R. R., 2004. Maximum-likelihood estimation of demographic parameters using the frequency spectrum of unlinked single-nucleotide polymorphisms. *Genetics* 168: 1699–1712.

*Beaumont M. A. and Rannala B., 2004. The Bayesian revolution in genetics. *Nature Reviews Genetics* 5: 251–261.

*Griffiths R. C. and Tavaré S., 1994. Ancestral Inference in Population Genetics. *Statistical Science* 9: 307–319.

He W., Lin L., Shen F., et al., 2008. Genetic diversities of the giant panda (*Ailuropoda melanoleuca*) in Wanglang and Baoxing Nature Reserves. *Conservation Genetics* 2008: 1541–1546.

*Hey J., 2010. The divergence of chimpanzee species and subspecies as revealed in multipopulation isolation-with-migration analyses. *Molecular Biology and Evolution* 27:921–933.

Krause J., Fu Q. M., Good J. M., et al., 2010. The complete mitochondrial DNA genome of an unknown hominin from southern Siberia. *Nature* 464: 894–897.

*Nielsen R. and Beaumont M. A., 2009. Statistical inferences in phylogeography. *Molecular Ecology* 18: 1034–1047.

*Nielsen R. and Wakeley J., 2001. Distinguishing migration from isolation: a Markov Chain Monte Carlo approach. *Genetics*. 158: 885–896.

*Novembre J., Johnson T., Bryc K., et al., 2008. Genes mirror geography within Europe. *Nature* 456:98–101.

Orti G., Bell M. A., Reimchen T. E. and Meyer A., 1994. Global survey of mitochondrial DNA sequences in the threespine stickle back: evidence for recent migrations. *Evolution* 48: 608–622.

*Rosenberg N. A., Pritchard J. K., Weber J. L., et al., 2002. The genetic structure of human populations. *Science* 298: 2381–2385.

Sokal R. R. and Michener C. D., 1958. A statistical method for evaluating systematic relationships. *University of Kansas Science Bulletin* 38: 1409–1438.

*Yi X., Liang Y., Huerta-Sanchez E., et al. (65 coauthors), 2010. Sequencing of 50 human exomes reveals adaptation to high altitude. *Science* 329: 75–78.
*Recommended reading

EXERCISES

5.1 A researcher obtains the following sample of genotypes from pandas living in two different geographic regions of China:

Genotype	AA	Aa	aa
Region 1	12	22	6
Region 2	32	6	2

Assume a simple divergence model in which the ancestral population size and the population sizes of both of the current populations are all equal to $2N = 10,000$. Furthermore, assume there has been no gene-flow since the time of divergence. Provide an estimate of the divergence time in number of years between the populations, assuming a generation time of one generation every five years.

5.2 Assuming an infinite sites model, mark the mutations in the sequences below on the tree as in Figure 5.2. If there are two or more possibilities for a mutation, mark all possible assignments on the tree. If the mutation in the site is not compatible with the tree under an infinite sites model, do not map the mutation on the tree.

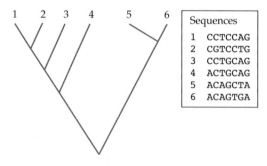

5.3 Answer the following questions about the trees in Figure 5.8:

  a. Which trees show evidence of reciprocal monophyly between population 1 and 2?

  b. If population 1 represents an African population and population 2 represents populations outside Africa, choosing between tree E and tree B, which tree appears most compatible with the out-of-Africa hypothesis and which tree appears most compatible with the multiregional hypothesis of human evolution?

c. Which trees are roughly compatible with the hypothesis that there has been no recent gene-flow between the populations?

5.4 Assume an infinite sites model, and a standard coalescence model of one population of constant size. Also assume that four mutations (SNPs) were observed in a data set of mtDNA sequences. Use Equation 3.19 (page 53) to plot the likelihood function for θ. Inspecting the plot, does it look as though the maximum likelihood estimate is close to Watterson's estimate of θ?

5.5 A panda bear has been rescued from a trap set by hunters. Now the question is: Which of the two regions from Exercise 5.1 does the panda bear originate from? The panda bear is genotyped for the locus discussed in Exercise 5.1 and is found to have genotype *aa*.

a. Based on this evidence alone, and assuming the prior probability of the panda bear being from either region is 0.5, what is the posterior probability that it is from region 1?

b. Now assume that it the panda bear was found much closer to region 2 than to region 1. You therefore assume that the prior probabilities from region 1 and 2 are 10% and 90%, respectively. Now what is the posterior probability that it is from population 1?

6 *Linkage Disequilibrium and Gene Mapping*

THE PREVIOUS CHAPTERS have treated each locus and each nucleotide position in isolation—as though they were all independent units. This approach creates a solid basis for studying population genetics, but it does not take into account that loci are arranged on chromosomes and hence are not inherited independently. Loci on each chromosome are linked and segregate together during meiosis. Considering two or more loci at a time is a challenge, because there are more things to keep track of, but it also provides an opportunity. Much can be learned from the study of linked loci, particularly if they are closely linked.

To motivate thinking about linked loci, look at the data represented in **Figure 6.1**. The figure shows sequences from several X chromosomes of West African human males. Polymorphic sites in a few genes on the same chromosome near the *G6PD* gene are shown. Blue shading indicates that the nucleotide is the same as in the human reference sequence, yellow that it is different. *G6PD* codes for an enzyme important for glycolysis. The *B* allele is functionally normal. There are several deficiency alleles that cause anemia. The A^+ allele differs from the reference sequence at sites 82, 87, and 91, and results in slightly lower enzymatic activity. The A^- allele, which has more reduced activity, has an A instead of a G at position 90, resulting in a methionine instead of a valine in the protein sequence. Females who are BA^- and males who are hemizygous for A^- are prone to developing hemolytic anemia when they suffer an infection or as a result of eating certain foods, notably fava beans.

Now look more closely at the figure and notice that in the *L1CAM* gene, which is roughly 700 kb away from *G6PD*, there are three sites at which nucleotides that are very rare on the A^+ and *B* chromosomes are found on a majority of the A^- chromosomes: fourteen out of twenty

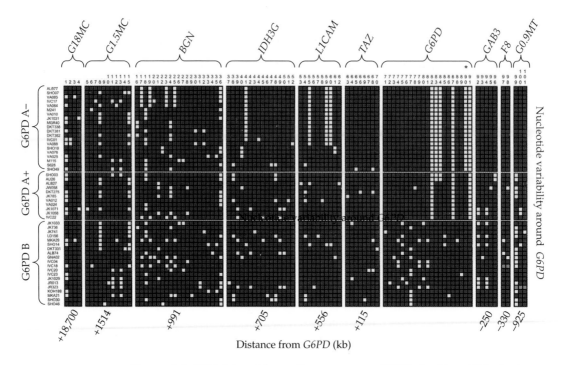

Figure 6.1 SNP data set for coding genes near *G6PD*. A⁻, A⁺ and B indicate functionally different alleles of *G6PD*. Blue indicates that the nucleotide is the same as the human reference sequence; yellow indicates the nucleotide differs from the reference sequence. The coding genes are identified by abbreviations at the top. The polymorphic sites are numbered across the top. The labels on the left are identifiers of the individuals whose X chromosomes were sequenced. (After Saunders et al., 2005.)

chromosomes share the same set of nucleotides at sites 55, 59, and 60. Clearly, there is some relationship between rare nucleotides in *L1CAM* and the A at site 90 in *G6PD*. If a chromosome has an A at site 90, then there is good chance it has the three unusual nucleotides in *L1CAM*. But if a chromosome has a G at site 90, then there is only a small chance of its having any of the rare nucleotides in *L1CAM*, let alone all three of them. We characterize this correspondence by saying that sites 55, 59, and 60 in *L1CAM* are in **linkage disequilibrium** (**LD**) with site 90 in *G6PD*. We will see that patterns of LD among closely linked loci can reveal surprising things about their recent history.

Linkage Disequilibrium

To introduce the concept of linkage disequilibrium, we start with the simplest case: two linked loci, each of which has two alleles—*A* and *a* at one locus and *B* and *b* at the other. The four possible combinations of alleles (*AB*, *Ab*, *aB*, and *ab*) are called **haplotypes**. In any sample of chromosomes, such as

---

BOX 6.1   Coefficients of Linkage Disequilibrium

Calculation of the coefficients of linkage disequilibrium (D) in a hypothetical example:

$$n_{AB} = 50, n_{Ab} = 20, n_{aB} = 10, n_{ab} = 20 \ (n = 100)$$

$$f_{AB} = 0.5, f_{Ab} = 0.2, f_{aB} = 0.1, f_{ab} = 0.2$$

$$f_A = f_{AB} + f_{Ab} = 0.7; f_B = f_{AB} + f_{aB} = 0.6; f_a = f_{aB} + f_{ab} = 0.3; f_b = f_{Ab} + f_{ab} = 0.4$$

$$D_{AB} = f_{AB} - f_A f_B = 0.5 - 0.42 = +0.08$$

$$D_{Ab} = f_{Ab} - f_A f_b = 0.2 - 0.28 = -0.08$$

$$D_{aB} = f_{aB} - f_a f_B = 0.1 - 0.18 = -0.08$$

$$D_{ab} = f_{ab} - f_a f_b = 0.2 - 0.12 = +0.08$$

---

those shown in Figure 6.1, we can count the numbers of each haplotype, n_{AB}, n_{Ab}, n_{aB}, and n_{ab}. Dividing by the total, we get the haplotype frequencies, f_{AB}, f_{Ab}, f_{aB}, and f_{ab}. Furthermore, we can get the frequencies of each allele by adding the appropriate haplotype frequencies. For example, the frequency of A is $f_A = f_{AB} + f_{Ab}$.

Alleles in the hypothetical example in **Box 6.1** are not independent. If there is an A on a chromosome, it is more likely that there is a B than a b at the other locus. To quantify how much more likely, we introduce the **coefficient of linkage disequilibrium, D.** If the presence of alleles at the two loci were independent of one another, the probability of any two occurring together would be the product of the allele frequencies.

For example, if A and B were independent, $f_{AB} = f_A f_B$. The coefficient of linkage disequilibrium between A and B, D_{AB}, is the difference between the actual haplotype frequency and the haplotype frequency calculated assuming independence:

$$D_{AB} = f_{AB} - f_A f_B \tag{6.1}$$

In Box 6.1, $D_{AB} = 0.08$.

If $D_{AB} = 0$, then $f_{AB} = f_A f_B$, and the alleles are said to be in **linkage equilibrium**. If $D_{AB} \neq 0$, they are in linkage disequilibrium. This terminology is somewhat unfortunate, because the word "disequilibrium" suggests a dynamic process, but that is not what it means in this context. D_{AB} is just a description of the relationship between haplotype and allele frequencies. It tells us nothing about whether the population sampled is at equilibrium.

Every pair of alleles has its own value of D:

$$D_{Ab} = f_{Ab} - f_A f_b$$
$$D_{aB} = f_{aB} - f_a f_B \qquad (6.2)$$
$$D_{ab} = f_{ab} - f_a f_b$$

When there are only two alleles at each locus, however, there is a simple relationship between the four coefficients of LD: $D_{Ab} = D_{aB}$, $D_{AB} = D_{ab}$, and $D_{AB} = -D_{Ab}$. We can see that numerically in Box 6.1. The algebraic proof of this fact is easily derived (see **Box 6.2**). There is only one numerical value, which we write as D without a subscript when there is no risk of ambiguity. The sign of D depends on which pair of alleles one starts with. If we start with A and B, $D = D_{AB} = D_{ab}$ and $D_{Ab} = D_{aB} = -D_{AB}$. If we start with A and b instead, the signs will be reversed. If there are more than two alleles at either or both loci, then more than one D is needed to fully characterize the extent of LD between two loci, and subscripts on D are necessary (See Exercise 6.3).

After computing D for a given data set, one question that arises is how to decide whether a given value of D is large or small. Suppose, for example, you are told that $D = 0.006$. Does this indicate only a slight degree of LD or a substantial amount? The answer depends on the allele frequencies, which constrain the range of values that D can possibly have. We can find those constraints by writing the haplotype frequencies in terms of D and the allele frequencies:

$$f_{AB} = f_A f_B + D$$
$$f_{Ab} = f_A f_b - D$$
$$f_{aB} = f_a f_B - D \qquad (6.3)$$
$$f_{ab} = f_a f_b + D$$

We know that none of the haplotype frequencies can be negative. Therefore, if $D > 0$, then D has to be less than the smaller of $f_A f_b$ and $f_a f_B$:

$$D \leq \min(f_A f_b, f_a f_B) \qquad (6.4)$$

BOX 6.2 LD Coefficients for Two Diallelic Loci

Recall that $D_{AB} = f_{AB} - f_A f_B$. By definition,

$$D_{Ab} = f_{Ab} - f_A f_b = f_A - f_{AB} - f_A f_b$$

where we have used the fact that $f_A = f_{AB} + f_{Ab}$. Collecting terms, we have $f_A - f_{AB} - f_A f_b = f_A(1 - f_b) - f_{AB} = f_A f_B - f_{AB} = -D_{AB}$, where we used the fact that $f_B + f_b = 1$. Similar calculations show that $D_{aB} = -D_{AB}$ and $D_{ab} = D_{AB}$.

If $D < 0$, then for the same reason,

$$-D \leq \min(f_A f_B, f_a f_b) \qquad (6.5)$$

Suppose first that $f_A = 0.01$ and $f_B = 0.4$. These inequalities tell us that $-0.004 \leq D \leq 0.006$. With these allele frequencies, $D = 0.006$ is the largest positive value D can have, so we would conclude that there is the maximum possible LD between these two loci. If, instead, $f_A = 0.3$ and $f_B = 0.4$, then $-0.12 \leq D \leq 0.18$ and $D = 0.006$ is only a small fraction of the maximum.

To describe the extent of LD in a way that takes the range of possible values into account, we define D' to be the ratio of $|D|$ to its maximum possible value:

$$D' = \frac{D}{\min(f_A f_b, f_a f_B)} \quad \text{if } D > 0$$
$$= \frac{-D}{\min(f_A f_B, f_a f_b)} \quad \text{if } D < 0 \qquad (6.6)$$

D' is guaranteed to be between 0 and 1.

D' is an especially useful description of LD, because if $D' = 1$, at least one haplotype is missing. The reason is that if $D' = 1$, D is at either its maximum positive value or its minimum negative value. In either case, Equation 6.3 shows that at least one of the haplotype frequencies has to be 0. For example if $D = 0.006$, $f_A = 0.01$, and $f_B = 0.4$, then $f_{Ab} = 0$. The reverse is also true. If only three of the four possible haplotypes are present, we can be sure that $D' = 1$. It is not necessary to do any calculations at all.

$D' = 1$ is an also an important special case for the following reason. Suppose that at some time, only the B/b locus is polymorphic. The other locus is fixed for a. Then in the next generation, a single copy of A appears due to mutation. In that generation, $D' = 1$. Why? Because only one chromosome carries the new A, and that chromosome carries either B or b. One of the haplotypes, either Ab or AB, has to be missing. Therefore, immediately after a mutation creates an allele not previously in the population, $D' = 1$ between that locus and every other polymorphic locus on the same chromosome, a fact that we will see has practical implications.

Another quantity that is widely used to characterize the extent of LD is

$$r^2 = \frac{D^2}{f_A f_a f_B f_b} \qquad (6.7)$$

The value of r^2 also is between 0 and 1. It is convenient for some purposes, because it is a correlation coefficient, as shown in **Box 6.3**, and because it is closely related to the χ^2 test of independence of the alleles at the two loci (**Box 6.4**). The χ^2 test was introduced in Box 1.3.

BOX 6.3 r^2 as a Correlation Coefficient

The term, r^2 defined by Equation 6.7, does not have the same intuitive appeal as D', but it is slightly easier to calculate and is convenient for statistical purposes because it is the square of a correlation coefficient. To see that, let each haplotype be represented by a pair of numbers i and j. Let $i = 1$ if the haplotype has an A and 0 if it has an a. And let $j = 1$ if the haplotype has a B and 0 if it has a b. Adding over all haplotypes:

$$Ave(i) = (1)\Pr(i = 1) + (0)\Pr(i = 0) = f_A$$

because f_A is the probability that a haplotype carries an A. In a similar way, $Ave(j) = f_B$. Furthermore, the variance of i depends on f_A:

$$Var(i) = Ave(i^2) - [Ave(i)]^2 = (1)\Pr(i = 1) + (0)\Pr(i = 0) - f_A^2 = f_A - f_A^2 = f_A f_a$$

In the same manner, $Var(j) = f_B f_b$. Ave indicates the average and Var indicates the variance.

The correlation coefficient between i and j is defined to be

$$r = \frac{Ave(ij) - Ave(i)Ave(j)}{\sqrt{Var(i)Var(j)}}$$

We can see that

$$Ave(ij) = (1 \times 1)f_{AB} + (1 \times 0)f_{Ab} + (0 \times 1)f_{aB} + (0 \times 0)f_{ab} = f_{AB}$$

Therefore,

$$r = \frac{f_{AB} - f_A f_B}{\sqrt{f_A f_a f_B f_b}}$$

and r^2 is given by Equation 6.7.

Evolution of D

As we have emphasized, D and its close relatives D' and r^2 are descriptions of haplotype frequencies in a population or in a sample from a population. If $D = 0$, then the presence of an allele at one locus in a haplotype is independent of the presence of an allele at the other locus. We next show that values of D can tell us something interesting because of the way that D changes when there is random mating. Assume that the allele frequencies and D are known. We know that in a large population, allele frequencies do not change after one generation of random mating. The question is: How does D change? A little algebra, outlined in **Box 6.5**, shows that

$$D(t + 1) = (1 - c)D(t) \tag{6.8}$$

where c is the recombination rate between the two loci. Recall from basic genetics that c is the fraction of gametes produced by a double heterozygote

---

BOX 6.4 r^2 and χ^2

We can represent the haplotype numbers in a sample of n chromosomes in a 2×2 table:

	A	a	Total B
B	nf_{AB}	nf_{aB}	nf_B
b	nf_{Ab}	nf_{ab}	nf_b
Total A	nf_A	nf_a	n

To do a χ^2 of independence of A and B from these data, we compute the expectations under independence and then add over all four haplotypes. The expectation in each cell is the product of the allele frequencies multiplied by the sample size. For example, the expected number of AB hapotypes is $nf_A f_B$. Therefore,

$$\chi^2 = \frac{n^2(f_{AB} - f_A f_B)^2}{nf_A f_B} + \frac{n^2(f_{Ab} - f_A f_b)^2}{nf_A f_b} + \frac{n^2(f_{aB} - f_a f_B)^2}{nf_a f_B} + \frac{n^2(f_{ab} - f_a f_b)^2}{nf_a f_b}$$

The numerator of each term is $n^2 D^2$ and the rest of the sum simplifies nicely:

$$\chi^2 = nD^2 \left(\frac{1}{f_A f_B} + \frac{1}{f_A f_b} + \frac{1}{f_a f_B} + \frac{1}{f_a f_b} \right) = \frac{nD^2}{f_A f_a f_B f_b}(f_a f_b + f_a f_B + f_A f_b + f_A f_B)$$

$$= \frac{nD^2}{f_A f_a f_B f_b} = nr^2$$

To test whether A and B are independent, we use a table of χ^2 values with 1 degree of freedom (df). There is 1 df because both f_A and f_B have to be estimated from the data.

(say an individual who carries one AB chromosome and one ab chromosome) that are recombinants (in this case, Ab and aB chromosomes).

To see why Equation 6.8 is so surprising, you also have to recall from basic genetics that $0 \leq c \leq \frac{1}{2}$. If $c = \frac{1}{2}$, then the two loci are inherited independently, conforming to Mendel's second law, which applies both to loci on different chromosomes and to loci far apart on the same chromosome. Equation 6.8 tells us that, even for loci on different chromosomes, D does not go to 0 after one generation of random mating. Its value is reduced by only a factor of $\frac{1}{2}$. That is very different from what happens to genotype frequencies at each locus. One generation of random mating is enough to create the Hardy–Weinberg frequencies at both loci, and that is true regardless of how closely linked they are. Linkage equilibrium takes longer—possibly much longer—to be established than Hardy–Weinberg equilibrium.

BOX 6.5 Change in *D* Due to Random Mating

To find the change in D after one generation of random mating, we have to consider the haplotypes produced by each individual, as shown in the table below. Note that it is necessary to distinguish individuals who are doubly heterozygous because they have one *AB* chromosome and one *ab* chromosome (denoted by *AB/ab*) from individuals who are doubly heterozygous because they have one *Ab* chromosome and one *aB* chromosome (*Ab/aB*). Random mating combines gametes produced by each individual and hence does not change haplotype frequencies. Therefore, to find the haplotype frequencies in the next generation, we add the contribution of each genotype.

Genotype	Frequency	Gametes Produced			
		AB	Ab	aB	ab
AB/AB	f_{AB}^2	1	0	0	0
AB/Ab	$2f_{AB}f_{Ab}$	½	½	0	0
AB/aB	$2f_{AB}f_{aB}$	½	0	½	0
AB/ab	$2f_{AB}f_{ab}$	$(1-c)/2$	$c/2$	$c/2$	$(1-c)/2$
Ab/Ab	f_{Ab}^2	0	1	0	0
Ab/aB	$2f_{Ab}f_{aB}$	$c/2$	$(1-c)/2$	$(1-c)/2$	$c/2$
Ab/ab	$2f_{Ab}f_{ab}$	0	½	0	½
aB/aB	f_{aB}^2	0	0	1	0
aB/ab	$2f_{aB}f_{ab}$	0	0	½	½
ab/ab	f_{ab}^2	0	0	0	1

For example,

$$f_{AB}(t+1) = f_{AB}^2 + 2f_{AB}f_{Ab}(\tfrac{1}{2}) + 2f_{AB}f_{aB}(\tfrac{1}{2}) + 2f_{AB}f_{ab}(1-c)/2 + 2f_{Ab}f_{aB}(c/2)$$
$$= f_{AB}^2 + f_{AB}f_{Ab} + f_{AB}f_{aB} + f_{AB}f_{ab} - c(f_{AB}f_{ab} - f_{Ab}f_{aB})$$
$$= f_{AB} - c(f_{AB}f_{ab} - f_{Ab}f_{aB})$$

We can simplify further by noting that

$$f_{AB}f_{ab} - f_{Ab}f_{aB} = (f_A f_B + D)(f_a f_b + D) - (f_A f_b - D)(f_a f_B - D)$$
$$= (f_A f_B + f_a f_b + f_A f_b + f_a f_B)D = D$$

Finally, we recall that $f_{AB} = f_A f_B + D$ and that one generation of random mating does not change allele frequencies at either locus. Therefore, we have

$$f_A f_B + D(t+1) = f_A f_B + D(t) - cD(t)$$

which reduces to Equation 6.8 in the main text.

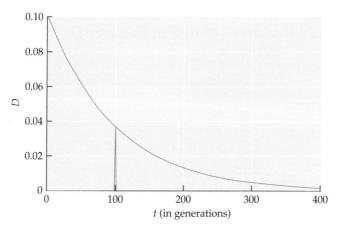

Figure 6.2 Decrease of D from an initial value of 0.1 when $c = 0.01$ (i.e., 1 cM). The vertical line shows that D decreases to about 37% of its initial value in $1/c$ generations.

Another important consequence of Equation 6.8 is that the rate of decrease of D depends on c. If c is small, D decreases only slightly in one generation of random mating. The effect becomes more pronounced after more generations. Applying Equation 6.8 repeatedly, we get

$$D(t) = (1 - c)^t\, D(0) \tag{6.9}$$

where $D(0)$ is the initial value of D. The utility of D and related quantities (D' and r^2) follows from this equation. Loci that are very tightly linked (the recombination rate c is small) will remain in LD for a long time, while loci that are less tightly linked (c is larger) will approach linkage equilibrium fairly quickly. After $1/c$ generations, D will decrease to roughly 37% of its initial value, as illustrated in **Figure 6.2**.

We can relate this result to the human genome, for which recombination rates are fairly well known. In humans, an approximate rule of thumb is that $c = 0.01$ (i.e., one centiMorgan or cM) corresponds roughly to 1 million bases or 1 mb. The average density of protein-coding genes is about 7 per mb (assuming 21,000 genes in 3,000 mb). Therefore protein-coding genes that are next to each other on a chromosome have an average recombination rate of approximately $c = 0.00142$. For that value of c, 1000 generations of random mating reduces D between adjacent genes to about 24% of the initial value. In humans, 1000 generations corresponds to about 25,000 years. Although all these numbers are only approximate, they give you the idea that any LD that existed between adjacent genes in the human genome early in the history of modern humans will probably have persisted until the present day.

BOX 6.6   Recurrent Mutation Reduces D'

Even in the absence of recombination, D' may be less than 1 if there is *recurrent* mutation—that is, if mutation from one allele to the other occurs more than once at either locus. If there is no recombination, both loci have the same gene genealogy, as illustrated in the figure for a sample of nine chromosomes.

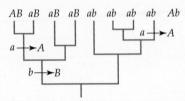

We now recall the point made about the special case of $D' = 1$. As we said, when a new mutation appears, $D' = 1$ because only three of the four possible haplotypes are present. D' decreases from 1 because of recombination: only recombination can create the fourth haplotype. If there is no recombination, then there will continue to be only three haplotypes, not four, and D' will remain 1. This fact provides a strong test for the occurrence of recombination, even if it occurs at a very low rate. If we see all four haplotypes, we know that some recombination must have occurred in the past, provided that we assume that each allele arises only once by mutation. If we assume that there is **recurrent mutation**—one allele arising repeatedly from independent mutations—then mutation alone may be responsible for the fourth haplotype (see **Box 6.6**).

We now return to the data presented in Figure 6.1. The fact that the mutation at site 90, which distinguishes the A^- allele, is in unusually strong LD with sites 750 kb away indicates that the mutation occurred in the recent past and for some reason increased in frequency. In fact, we know that $A-$ increased in frequency because it provides some protection against malaria, in a way similar to the S allele of β-*globin* gene. As it increased in frequency, it carried with it closely linked markers that remain in strong LD because recombination has not had time to break it down. This process is called *genetic hitchhiking* and will be discussed in more detail in Chapter 8.

Two-Locus Wahlund Effect

The Wahlund effect, introduced in Chapter 4, is a decrease in heterozygosity in subdivided populations below what is expected from the Hardy–Weinberg frequencies. When two loci are considered at a time, population subdivision creates linkage disequilibrium, even if there is no LD in each subpopulation. This is called the **two-locus Wahlund effect**. To see this, consider the following simple situation. Assume there are two subpopulations, one fixed for A and B and the other fixed for a and b. You mistakenly think that there is only one population and mix samples from the two subpopulations together. You determine genotypes of individuals in order to find the allele and haplotype frequencies. Suppose half of your sample is from the AB population and half from the ab population. What would you conclude?

Half of your sample would consist of $AABB$ individuals and half of $aabb$ individuals. You would find $f_A = f_B = \frac{1}{2}$ and $f_{AB} = f_{ab} = \frac{1}{2}$, and conclude that $D = \frac{1}{4}$. The apparent LD in your sample is there because you mixed individuals from two populations that differ in allele frequency. Neither population is polymorphic, so $D = 0$ in both. This is an extreme example of the two-locus Wahlund effect. The mixing of individuals from populations with different allele frequencies creates a positive inbreeding coefficient. The two-locus Wahlund effect is similar. Mixing of individuals from populations in which allele frequencies differ at both loci will create LD between them, even if both populations separately are in linkage equilibrium. The general result is derived in **Box 6.7**.

BOX 6.7 Two-Locus Wahlund Effect

We will show algebraically that differences in allele frequencies among subpopulations create linkage disequilibrium when chromosomes are sampled from different subpopulations. For simplicity, we will consider the case of two subpopulations and assume that equal numbers of chromosomes are sampled from each. Assume there are two alleles at each of two loci (A and a at one locus and B and b at the other) and that within each subpopulation, the two loci are in linkage equilibrium.

Because there is linkage equilibrium in each subpopulation, the frequencies of the AB haplotype in the two subpopulations are $f_A^{(1)} f_B^{(1)}$ and $f_A^{(2)} f_B^{(2)}$, where the superscript indicates the subpopulation. The frequency of AB in a mixture chosen from the two subpopulations is the average:

$$f_{AB} = \frac{1}{2} f_A^{(1)} f_B^{(1)} + \frac{1}{2} f_A^{(2)} f_B^{(2)}$$

Furthermore, the allele frequencies in the mixture are also the averages:

$$f_A = \frac{1}{2} f_A^{(1)} + \frac{1}{2} f_A^{(2)}; \quad f_B = \frac{1}{2} f_B^{(1)} + \frac{1}{2} f_B^{(2)}$$

The coefficient of LD in the mixture is

$$D = f_{AB} - f_A f_B =$$
$$= \frac{1}{2} f_A^{(1)} f_B^{(1)} + \frac{1}{2} f_A^{(2)} f_B^{(2)} - \left(\frac{1}{2} f_A^{(1)} + \frac{1}{2} f_A^{(2)} \right)\left(\frac{1}{2} f_B^{(1)} + \frac{1}{2} f_B^{(2)} \right)$$
$$= \frac{1}{4} \left(f_A^{(1)} - f_A^{(2)} \right)\left(f_B^{(1)} - f_B^{(2)} \right)$$

We can see that if the allele frequencies in the subpopulations are different for both loci, D will be non-zero in the mixture.

If only a one pair of loci is sampled, it may be difficult to know whether the LD detected is the result of the two-locus Wahlund effect or of some other process such as natural selection or genotyping error. If many more loci are genotyped and if most pairs of loci are in significant LD, you would suspect that the two-locus Wahlund effect is responsible. Mixing of individuals from different subpopulations creates LD whenever allele frequencies at both loci differ among subpopulations. However, if only one pair of loci is in strong LD and the rest are in linkage equilibrium, you would suspect that selection or error is the cause.

Genealogical Interpretation of LD

In Chapter 3, you were introduced to the idea of a gene genealogy for a single locus. Each locus has a history represented by a gene genealogy that indicates the times in the past when their ancestral lineages coalesced. For two linked loci, there are two gene genealogies. The relationship between these two gene genealogies depends on the history of recombination between the loci.

Recombination

We can start by imagining the ancestry of two loci on a single chromosome sampled today. If there is no recombination between them ($c = 0$), then the ancestors of both loci were on the same chromosome every generation in the past. The reason is that, in the absence of recombination, the chromosome was replicated during meiosis and transmitted intact to each descendant. That is the situation with the mitochondrial chromosome and Y chromosome in humans and other mammals. As a consequence, the gene genealogies of the two loci are identical, and ancestry of both loci are represented by a single line.

If there has been recombination ($c > 0$), the genealogies of the two loci are different. Looking backwards in time, one of two things happened in each generation. Either there was no recombination between the two loci, or there was recombination between them. The two possibilities are illustrated in **Figure 6.3**. If there was no recombination (top part), the two loci remain on the same chromosome. If there was recombination (bottom part), then the ancestral lineages were on different chromosomes the generation before and are joined by recombination in that generation. Starting at the present and going

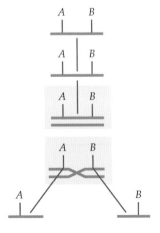

Figure 6.3 Recombination separates gene genealogies. In the top part, there is no recombination in the previous generation, so the ancestral lineages are on the same chromosome. In the bottom part, there is recombination The ancestors of the ancestral lineages were on different chromosomes in the previous generation and joined by crossing over during meiosis.

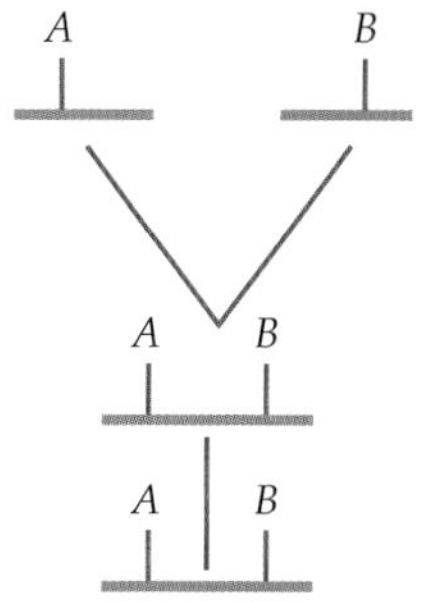

Figure 6.4 Coalescence joins gene genealogies. When coalescence occurs, the chromosomes carrying the *A* and *B* lineages are descended from a single ancestral chromosome.

backward in time, eventually recombination between the two loci will occur in an ancestor and before that time, the gene genealogies will be different.

When the ancestral lineages of the two loci are on different chromosomes, they remain on different chromosomes until those two chromosomes coalesce, as illustrated in **Figure 6.4**. Before that coalescence, the two lineages once again share the same chromosome, and they will remain together on one chromosome until they are again separated by recombination in an earlier ancestor.

Over a long time, it is as if the two ancestral lineages are doing a simple dance. Some of the time they are together, and some of the time they are apart. The relative amount of time in each condition depends on the recombination rate and the population size. When there is a single lineage, recombination splits that lineage at a rate c per generation. The average waiting time until the split is $1/c$ generations. When there are two lineages, they will coalesce with probability $1/(2N)$ per generation. The average waiting time until they join is $2N$ generations. If c is much smaller than $1/(2N)$, then most of the time there will be a single ancestral lineage. If c is much larger than $1/(2N)$, then most of the time there will be two ancestral lineages.

Next consider a pair of chromosomes sampled today. The two chromosomes might have been produced by coalescence in an earlier generation, which would mean that the gene genealogies of two loci on the chromosome are identical. Or one of the chromosomes might have been produced by recombination in an ancestor, which would mean that the two loci have different gene genealogies. The situation is very much like the appearance of mutation in a gene genealogy (Example 8 in Appendix A). The problem is to find the probability that coalescence occurs before recombination affects either lineage. The probability per generation of a coalescence is $1/(2N)$ and the probability per generation that there is no recombination in either lineage is $(1-c)^2$. Therefore, the probability that there is no recombination before coalescence is approximately

$$\frac{1/(2N)}{1/(2N)+2c} = \frac{1}{1+4Nc} \tag{6.10}$$

when N is large and c is small (see pp. 241–243 of Appendix A). As in the case of mutation and coalescence, a combination of parameters $4Nc$ indicates the relative importance of recombination and coalescence. When $4Nc$ is much less than 1, recombination occurs much less often than coalescence, and we expect the gene genealogies of linked loci to be the nearly the same. When $4Nc$ is much greater than 1, recombination occurs much more often than coalescence, and we expect the gene genealogies of the two loci to be quite different.

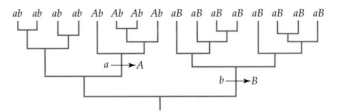

Figure 6.5 Three haplotypes created by single mutations when two loci have the same gene tree.

There is a close relationship between the gene genealogies of two loci and their coefficient of linkage disequilibrium. Suppose that at each locus, there was a single mutation that created the second allele. If the two loci have the same gene genealogy (**Figure 6.5**), then only three of the four possible haplotypes can be present. The actual haplotype and allele frequencies and hence D depend on where in the gene genealogy the two mutations occurred. But we can be sure that $D' = 1$. Only if the gene genealogies are not the same or if there was recurrent mutation can D' be less than 1. The less similar the two gene genealogies are, the smaller D' will be.

If we consider now not two loci but a whole segment of a chromosome, a reasonably simple picture emerges. Sites that are very closely linked probably have the same or nearly the same gene genealogies. They are likely to have large D' and r^2 values. Sites that are farther apart have higher recombination rates between them and would be expected to have less similar gene genealogies and hence smaller D' and r^2. And sites that are very far apart should have quite different gene genealogies, implying that D' that is 0 or nearly so. That is, we would expect to see a decrease in the extent of LD between pairs of sites separated by increasing distances on a chromosome. That expectation is seen on average in data from humans and other species but with enormous variation, as shown in **Figure 6.6** for loci separated by different distances. The difference in D' between sites separated by the same distance on the chromosome results from the basic unpredictability of coalescence and recombination. The actual extent of similarity of two gene genealogies of linked sites depends on exactly when in the past recombination affected those genealogies. Only a few recombination events in the history of two loci can affect the gene genealogies of closely linked sites and exactly how many occur cannot be predicted even when the recombination rates are perfectly known.

We contrast Figure 6.6 with **Figure 6.7**, plotting D' against distance on the mitochondrial chromosome. In this case, there is no decrease of D' with increasing distance, which is what we expect, because there is no recombination in human mtDNA. Note that $D' \neq 1$ for all pairs of sites, however. This indicates that there was recurrent mutation at many sites, which is consistent with the fact that the mutation rate in mtDNA is much higher than that in nuclear DNA.

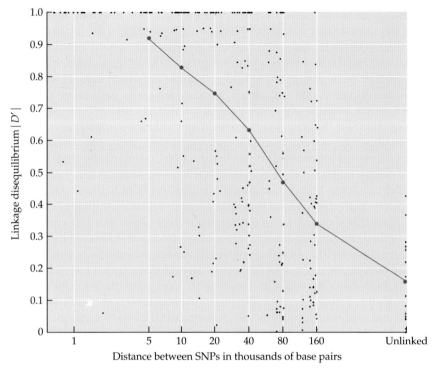

Figure 6.6 *D′* plotted against distance separating SNPs in the human genome. The dots indicate the average values of *D′*. (After Reich et al., 2001.)

Association Mapping

One of the goals of modern genetics is to map genes that affect important phenotypic traits. By mapping a gene, we mean determining the recombination rates between the gene of interest and marker loci whose positions *in genome already known*

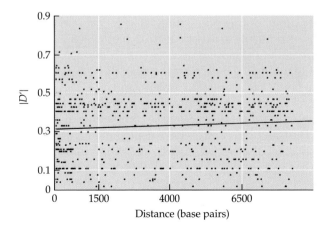

Figure 6.7 *D′* versus distance in the mtDNA genome of non-African individuals. The line is fitted to the averages at each distance. (After Ingman et al., 2001.)

in the genome are already known. For example, we might be able to say that the recombination rate between a marker locus and the locus causing a particular genetic disease in humans is 0.001. That would tell us that the locus causing the disease is within roughly 100 kb of the marker, but without further information we would not know on which side. If we knew the recombination rate with other markers, we could get a more accurate estimate of where the unknown gene of interest is. Linkage disequilibrium is important for the success of genomic methods for mapping genes.

To illustrate, we will briefly describe efforts to map genes affecting the risk of complex inherited diseases in humans. Complex inherited diseases are diseases that tend to be concentrated in some families but that are not caused by single Mendelian genes. Various forms of heart disease, most cancers, schizophrenia, and multiple sclerosis are all examples of complex inherited diseases. Evidence for genetic effects comes from the increased risk of these diseases in close relatives of affected individuals. For example, Crohn's disease (CD), an inflammatory disease of the intestines, tends to run in families. In one study, 8 of 18 pairs of monozygotic (genetically identical) twins, both had CD, while in the other 10 pairs, only one had the disease. Of 26 pairs of dizygotic (nonidentical) twins of whom at least 1 twin had CD, there was only one pair in which both had CD. The fact that the disease did not affect all of the genetically identical twins indicates that CD is not caused only by genetic effects. The environment and random events during development also play an important role in the development of the disease.

Another way to characterize the importance of genetic effects is to quantify the proportional increase in risk to close relatives of an affected individual. The increase in risk is called the *risk ratio*, λ_R, where R indicates the relationship (e.g., full sibling, first cousin, etc.). The average occurrence of CD is about 1 in 10,000. But an individual whose full sibling has CD has a 20 in 10,000 chance of also having the disease, so for this individual, $\lambda_{FS} = 20$ (FS indicates *full sibling*). Other complex diseases have smaller risk ratios. For example, $\lambda_{FS} \approx 15$ for type 1 diabetes, $\lambda_{FS} \approx 10$ for schizophrenia, $\lambda_{FS} \approx 3$ for type 2 diabetes, and $\lambda_{FS} \approx 2$ for most common cancers. As a practical matter, risk ratios are only approximate, because they are difficult to estimate, and estimates typically differ among populations. The increase in disease risk in close relatives is attributable to several loci, each of which increases disease risk by a small amount. Mapping loci that affect disease risk is a first step toward understanding the biological causes of these diseases and ultimately may lead to better prevention, treatment, and cure.

The starting point for mapping genes affecting a complex disease is a group of affected individuals (*cases*) and a group of comparable unaffected individuals (*controls*). The cases and controls have to be similar with respect to other factors that can affect disease risk, such as age, ethnic background, and economic status. For each marker locus in the study, the numbers of copies of one allele in the cases (n_{case}) and controls ($n_{control}$) are recorded. Then a χ^2 or other statistical test is performed on the data for each marker to determine whether a marker allele is significantly more common in

---

BOX 6.8   Example of a Case-Control Test

As an illustration, we use the test for association between a SNP on chromosome 16 in humans (*rs12708716*) and type 1 diabetes (T1D). The numbers are based on information from the Wellcome Trust case-control study.

	Cases	Controls
A	1300	2109
G	700	891

For this table, $\chi^2 = 15.53$, which implies $P < 8 \times 10^{-5}$.

---

cases than in controls. **Box 6.8** gives an example. If a significant difference is found, it is likely that the marker locus is closely linked to the locus that actually contributes to disease risk.

In most case-control studies today, very large numbers of marker loci, 500,000 or more, are surveyed. Such studies are called **genome-wide association studies** (**GWAS**) because they test for significant associations with marker loci throughout the genome. In a GWAS with 500,000 SNPs in a genome with 3×10^9 bases, the average spacing of SNPs is 6 kb. As shown in Figure 6.6, that spacing is close enough that any causative locus will likely be in LD with at least one of the nearby marker loci.

An important practical problem arises because a single GWAS requires as many χ^2 tests of significance as there are SNPs tested. Although the calculations are easy to do using efficient computer programs, the interpretation of the results requires some additional thought. Because so many tests are performed, many of them will appear to give significant results even for markers nowhere near loci that affect disease risk. For example, if we were to test 500,000 SNPs and say that there is a significant association between a SNP and the disease if the P value from a χ^2 test is less than 0.001, we would expect to see 500 SNPs significantly associated with higher risk even if there were no genetic effect at all on the disease. The solution is relatively simple: a much lower P value is used, typically 10^{-7} or smaller. But there is a cost to adopting that solution. Using a lower P value means that some loci that are actually associated with higher risk might be missed.

A study carried out by the Wellcome Trust illustrates many of the features of GWAS. A large consortium of medical doctors, geneticists, and statisticians studied seven complex diseases: bipolar disorder, coronary artery disease, Crohn's disease, rheumatoid arthritis, hypertension, type 1 diabetes, and type 2 diabetes. For each disease, approximately 2000 cases were genotyped at 500,000 SNPs. One set of 3000 individuals unaffected by any of the diseases was used as controls for each of the seven diseases. These "shared" controls reduced the total number of individuals that had to be genotyped. The results, some of which are presented in **Table 6.1**,

TABLE 6.1   Regions of the genome showing the strongest association signals

Collection	Chromosome	Genotypic P value	Risk allele	Minor allele
BD	16p12	$6.29 3 10^{-08}$	A	G
CAD	9p21	$1.16 3 10^{-13}$	C	C
CD	1p31	$5.85 3 10^{-12}$	T	T
CD	2q37	$5.26 3 10^{-14}$	T	C
CD	3p21	$3.58 3 10^{-08}$	A	A
CD	5p13	$1.99 3 10^{-12}$	G	G
CD	5q33	$3.15 3 10^{-07}$	T	T
CD	10q21	$1.75 3 10^{-06}$	G	A
CD	10q24	5.82×10^{-08}	G	G
CD	16q12	$3.98 3 10^{-11}$	G	G
CD	18p11	$2.03 3 10^{-07}$	G	G
RA	1p13 1	$5.55 3 10^{-25}$	A	A
RA	6	$5.18 3 10^{-75}$	T	T
T1D	1p13	$5.43 3 10^{-26}$	A	A
T1D	6	$5.47 3 10^{-134}$	A	G
T1D	12q13	$9.71 3 10^{-11}$	C	C
T1D	12q24	$1.51 3 10^{-14}$	G	G
T1D	16p13	$4.92 3 10^{-07}$	A	G
T2D	6p22	$3.34 3 10^{-07}$	C	C
T2D	10q25	$5.05 3 10^{-12}$	T	T
T2D	16q12	$1.91 3 10^{-07}$	A	A

Note: Regions with at least one SNP with a *P* value of less than 5×10^{-7} in the primary analyses. The minor allele is defined in the controls and its frequency in that group as well as the case sample is reported. SNPs significantly associated with the risk of one of six complex diseases. BD = Bipolar disorder; CAD = Coronary artery disease; CD = Crohn's disease; RA = Rheumatoid arthritis; T1D = Type 1 diabetes; T2D = Type 2 diabetes.

are typical of those from many other GWAS. Some features to note are: (1) alleles significantly associated with a disease are in moderate frequency ($0.2 < f < 0.8$); (2) each allele has only a modest effect on disease risk, increasing risk by a factor of 1.2 to 2; and (3) the success rate varies among diseases—no significant associations were found for hypertension, only one was found for bipolar disorder and coronary heart disease, but seven were found for type 1 diabetes and nine for Crohn's disease.

TABLE 6.1 (*continued*)

Heterozygote odds ratio	Homozygote odds ratio	Control MAF	Case MAF
2.08 (1.60–2.71)	2.07 (1.6–2.69)	0.282	0.248
1.47 (1.27–1.70)	1.9 (1.61–2.24)	0.474	0.554
1.39 (1.22–1.58)	1.86 (1.54–2.24)	0.317	0.391
1.19 (1.01–1.41)	1.85 (1.56–2.21)	0.481	0.402
1.09 (0.96–1.24)	1.84 (1.49–2.26)	0.282	0.331
1.54 (1.34–1.76)	2.32 (1.59–3.39)	0.125	0.181
1.54 (1.31–1.82)	1.92 (0.92–4.00)	0.067	0.098
1.23 (1.05–1.45)	1.55 (1.3–1.84)	0.461	0.406
1.2 (1.03–1.39)	1.62 (1.37–1.92)	0.477	0.537
1.29 (1.13–1.46)	1.92 (1.58–2.34)	0.287	0.356
1.3 (1.14–1.48)	2.01 (1.46–2.76)	0.163	0.208
1.98 (1.72–2.27)	3.32 (1.93–5.69)	0.096	0.168
2.36 (1.97–2.84)	5.21 (4.31–6.30)	0.489	0.685
1.82 (1.59–2.09)	5.19 (3.15–8.55)	0.096	0.169
5.49 (4.83–6.24)	18.52 (27.03–12.69)	0.387	0.150
1.34 (1.17–1.54)	1.75 (1.48–2.06)	0.423	0.493
1.34 (1.16–1.53)	1.94 (1.65–2.29)	0.424	0.506
1.19 (0.97–1.45)	1.55 (1.27–1.89)	0.350	0.297
1.18 (1.04–1.34)	2.17 (1.6–2.95)	0.178	0.218
1.36 (1.2–1.54)	1.88 (1.56–2.27)	0.324	0.395
1.34 (1.17–1.52)	1.55 (1.3–1.84)	0.398	0.453

Note: No associations were found for hypertension. MAF = minor allele frequency, which is the frequency of the less common allele. The odds ratio is roughly the factor by which the disease risk is increased in an individual with that genotype.

Source: Wellcome Trust Case Control Consortium (2007).

Some of these results were not surprising. Relatively common alleles were detected because GWA studies are best suited for finding common alleles that affect disease risk and are ineffective at finding alleles in very low or high frequency. Alleles with only a modest effect on risk were detected because alleles of much larger effect would have already been detected by other means. What is surprising is that so few alleles were found, considering the size of the study, its cost, and the great effort that went into it. Other, more recent studies have found additional SNPs associated with each of these diseases, but it is becoming clear that many alleles associated with increased disease risk have too small an effect on average risk to be detectable in GWAS even with very large sample sizes. The variation in the success rate for different diseases was also somewhat surprising. Some diseases will yield their genetic secrets more readily than will others.

References

Ingman M., Kaessmann H., Pääbo S., Gyllensten U., 2000. Mitochondrial genome variation and the origin of modern humans. *Nature* 408: 708–713. http://www.nature.com/nature/journal/v408/n6813/abs/408708a0.html

Reich D. E., Cargill M., Bolk S., et al., 2001. Linkage disequilibrium in the human genome. *Nature* (London) 411: 199–204. http://www.nature.com/nature/journal/v411/n6834/abs/411199a0.html

*Satsangi J., Jewell D. P., Bell J. I., 1997. The genetics of inflammatory bowel disease. *Gut* 40: 572–574. http://gut.bmj.com/content/40/5.toc

Saunders M. A., Slatkin M., Garner C., et al., 2005. The span of linkage disequilibrium caused by selection on *G6PD* in humans. *Genetics* 171: 1219–1229. http://www.genetics.org/cgi/reprint/171/3/1219

*Wellcome Trust Case Control Consortium (2007) Genome-wide association study of 14,000 cases of seven common diseases and 3,000 shared controls. *Nature* 447: 661–678. http://www.nature.com/nature/journal/v447/n7145/abs/nature05911.html

*Recommended reading

EXERCISES

6.1 Calculate D, D' and r^2 for sites 82 and 83 in the data shown in Figure 6.1.

6.2 For a locus with two alleles, what are the constraints on the allele frequencies if two of the four possible haplotypes are missing?

6.3 If there are more than two alleles per locus, more coefficients of LD are needed to fully characterize the data. For the table below, calculate the six possible coefficients of LD: $D_{ij} = f_{A_i B_j} - f_{A_i} f_{B_j}$.

	A_1	A_2	A_3	Total
B_1	12	28	0	40
B_2	18	22	20	60
Total	30	50	20	100

6.4 The gene genealogy below represents the history of sample of a non-recombining segment containing two loci. The arrows indicate the lineages on which mutations occurred.

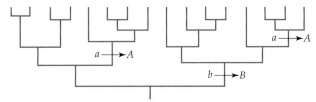

  a. What are the haplotypes at each tip of the gene genealogy if the ancestral chromosome carried a and b?

  b. Find D and D' for the sample.

6.5 Suppose you are studying the inheritance of two autosomal genes on the same chromosome that are sufficiently far apart that $c = \frac{1}{2}$. You begin with two populations, one of which is homozygous for A and B and the other of which is homozygous for a and b. You form the F_1 population by hybridizing the two populations. You then let the members of the F_1 population randomly mate to form the F_2 population. (This experimental design should look familiar. It is Mendel's design for testing for independent assortment. The only difference is that Mendel self-fertilized the plants in the F_1 rather than let them mate randomly.)

  a. What is D in the F_1 population?

  b. What is D in the F_2?

  c. Why does the correct answer to part a not agree with the prediction of Equation 6.8?

6.6 Suppose you sample haplotypes from a population and find the following counts: AB: 30, Ab: 270, aB: 370, and ab: 330?

  a. Use a χ^2 test to determine whether there is significant linkage disequilibrium at the 1% level or less in this population. ($P < 0.01$ if $\chi^2 > 6.636$ with 1 degree of freedom, which is appropriate for this test.)

  b. What is D in this sample?

  c. With these allele frequencies, what is the maximum absolute value of D if the two loci are not in significant LD at the 1% level. (Hint: Use the formula for χ^2 as a function of D and the allele frequencies, given in Box 6.4.)

  d. Assume that the recombination rate c is 0.001. Using Equation 6.8, determine how many generations of random mating will be necessary before there is no longer significant LD at the 1% level between these two loci. Assume that the allele frequencies do not change.

6.7 Suppose that you sample chromosomes from two populations and determine the haplotype frequencies in each. The data are shown in the table below:

	n_{AB}	n_{Ab}	n_{aB}	n_{ab}
Population 1	70	0	10	20
Population 2	20	10	0	70

a. What are the coefficients of LD in each population?

b. If the samples from the two populations were mixed, what would be the additional LD created by the two-locus Wahlund effect?

c. Is D' larger or smaller in the mixture than in the two populations separately?

6.8 Suppose you are interested in the gene genealogies of two loci with a recombination rate $c = 0.001$ between them.

a. For a single chromosome sampled from a randomly mating population containing N diploid individuals, what is the average number of generations in the past before the ancestral lineages of the two loci are on different chromosomes?

b. Why does the answer to part a not depend on N?

c. Once the ancestral lineages are on different chromosomes, what is the average time until they are again on one chromosome if $N = 100$ and $N = 1,000,000$?

d. Why does the answer to part c not depend on c?

e. What is the average time the two ancestral lineages are on one chromosome if $N = 100$ and $N = 1,000,000$?

6.9 Suppose you conduct a case-control study for the association between a SNP and the risk of type 2 diabetes and you find the following results:

	Cases	Controls
A	650	550
G	350	450

Is there a significant association between this SNP and the risk of type 2 diabetes?

7 *Selection I*

SO FAR, WE HAVE IGNORED THE POSSIBILITY that different alleles at a locus may affect survival and reproduction. That has allowed us to understand the consequences of random mating, genetic drift, and recombination. In this and the following three chapters, we will show how to allow for differences in survival and reproduction caused by differences in the genotype. In this chapter we introduce the principles of selection. We will start with selection on haploid organisms, which is relatively simple, but nonetheless reveals many of the important features of selection in diploid organisms. Then we discuss *viability selection* in diploids, which is selection related to differences in the chance of surviving from the zygote stage to the adult stage. At the end of the chapter, we will discuss *fertility selection*, selection resulting from the incompatibility of mating pairs. In Chapter 8, we will present the interaction of selection with genetic drift and mutation. In Chapter 9, we will summarize several methods that are used to detect the effect of natural selection in the genome. In Chapter 10, we will describe more complicated kinds of selection, including kin selection and genomic conflict, which involve interactions among individuals or interactions within the genome.

Selection in Haploids

We can illustrate many of the important ideas about how selection changes allele frequencies by considering haploid organisms that reproduce by binary fission. An example is of the adaptation of the bacteriophage MS2, which was exposed to elevated temperatures for a prolonged period. In several independent replicate lines, the mutation C206U increased in frequency as shown in **Figure 7.1**. Although there is some variation among replicates indicated by the error bars at each time point, on average, the frequency of this mutation increased steadily after the population was

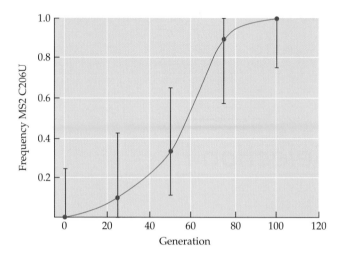

Figure 7.1 Time-series data for the mutation at the *C206U* locus in the bacteriophage MS2. Binomial confidence intervals are shown as black bars. (After Bollback and Huelsenbeck, 2007.)

exposed to higher temperatures, presumably because the mutation creates some advantage in growth and reproduction.

We can describe the increase in frequency of an allele that confers some advantage. To begin, we assume that at some time ($t = 0$) we have a population made up of two types, those that carry allele A at a locus and those that carry a. Let N_A and N_a be the numbers of the two types. The frequency of A is the fraction of the population that carries A: $f_A = N_A / (N_A + N_a)$. Now suppose that all the individuals have the opportunity to divide and form the next generation. Whether an individual survives and divides or not depends in part on its genotype at the A/a locus. In the next generation ($t = 1$), each A-bearing individual has on average w_A descendants, and each a-bearing individual has on average w_a descendants. The w's depend on both the rate of division of each type and the probability that they will survive long enough to divide. What we care about is the average number of A-bearing individuals in the next generation produced by each A-bearing individual in the current generation. A difference in reproductive rate indicates that one type survives more readily and divides more quickly than the other under the same conditions. The values of w depend on the current environmental conditions and will probably change if the environment changes. For example, *E. coli* and other bacteria have alleles that confer resistance to antibiotics. In the presence of antibiotics, bacteria carrying resistance alleles have a much higher growth rate, but in the absence of an antibiotic, the bacteria divide at a normal rate and the resistance allele remains at the same frequency unless it interferes with survival and cell division. For our purposes, the biological reason for the difference in growth rate is unimportant. What is important is that the reproductive rates differ, on average, because individuals have an A or a allele at this particular locus.

At $t = 1$, there will be $w_A N_A$ A-bearing individuals and $w_a N_a$ a-bearing individuals. The allele frequency will have changed from $f_A(0) = N_A/(N_A + N_a)$ to

$$f_A(1) = \frac{w_A N_A}{w_A N_A + w_a N_a} \tag{7.1}$$

If we are concerned only with changes in allele frequency, the absolute numbers of the two types do not matter. We can see this by dividing the numerator and denominator by $N_A + N_a$:

$$f_A(1) = \frac{w_A f_A(0)}{w_A f_A(0) + w_a f_a(0)} \tag{7.2}$$

We can illustrate this result with a numerical example. Suppose that initially, $N_A = 1000$ and $N_a = 3000$ ($f_A(0) = \frac{1}{4}$), and suppose also that $w_a = 1.5$ and $w_A = 1.7$. Equation 7.2 tells us that $f_A(1) = 0.25 \times 1.7/(0.25 \times 1.7 + 0.75 \times 1.5) = 0.274$. The frequency of A increases, because A-bearing individuals have more offspring than a-bearing ones.

Equation 7.2 also tells us that it is not the absolute reproductive rates (the w's) that determine the new allele frequency, but only the relative rates. We can see this in our numerical example, by letting $w_a = 17$ and $w_A = 15$. The population size at $t = 1$ will be 10 times as large, but f_A will still be 0.274. The same is true if the reproductive rates are $\frac{1}{10}$ larger ($w_a = 0.15$ and $w_A = 0.17$). Now the population size decreases, but the fraction of the population that carries an A still increases.

We can see algebraically that only the relative reproductive rates are important by dividing the numerator and denominator of Equation 7.2 by w_A to get

$$f_A(t+1) = \frac{f_A(0)}{f_A(0) + (w_a/w_A) f_a(0)} \tag{7.3}$$

The new frequency depends on the ratio w_a/w_A. It is convenient to define the *selection coefficient*, s, in terms of this ratio:

$$\frac{w_a}{w_A} = 1 - s \tag{7.4}$$

The selection coefficient summarizes the difference in reproductive rates in a way that makes it easy to calculate the effect on allele frequency:

$$f_A(1) = \frac{N_A}{N_A + (1-s)N_a} \tag{7.5}$$

Now, suppose that the difference in reproductive rate persists for a long time. After t time steps, $N_A(t) = w_A^t N_A$ and $N_a(t) = w_a^t N_a$. Both types in the

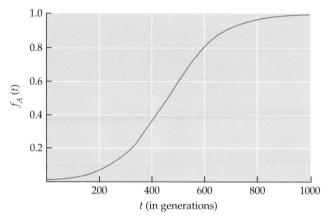

Figure 7.2 Graph of $f_A(t)$ for $f_A(0) = 0.01$ and $s = 0.01$.

population are growing (or declining) exponentially, but with different growth rates. Dividing by the total, we find:

$$f_A(t) = \frac{w_A^t f_A(0)}{w_A^t f_A(0) + w_a^t f_a(0)} = \frac{f_A(0)}{f_A(0) + (1-s)^t f_a(0)} \tag{7.6}$$

This equation leads to a simple prediction, illustrated in **Figure 7.2**, which is similar to the average curve shown in Figure 7.1. If $s > 0$, which means that A-bearing individuals reproduce more rapidly, f_A will increase from its initial value to 1. If f_A is small at first, the curve is a *sigmoid* curve, because it looks like an S tilted sideways. f_A increases slowly at first, then more rapidly when f_A is moderate, and then slowly again when f_A approaches 1. Eventually f_A will reach 1, meaning that A has been fixed by natural selection and a has been lost.

In Equation 7.6, $f_A(t)$ depends on the initial allele frequency and on $(1-s)^t$. Because $(1-s)^t$ is very close in value to e^{-st}, we can see that the allele frequency at any time depends on the product st. If s is increased by a factor of 10, for instance, then the time needed for f_A to reach a given value is decreased by a factor of 10. Roughly speaking, the time it takes for allele frequency to change substantially is $1/s$. In **Box 7.1**, we solve Equation 7.6 for t to determine how much time is needed for selection to cause a given change in frequency.

If s is negative, then a-bearing individuals have the advantage. Equation 7.6 implies that f_A will go to 0 as t becomes large. The formulas in Box 7.1 can be used in that case as well.

Selection in Diploids

Selection in diploids is similar in many ways to selection in haploids, but with the additional complication that diploid individuals do not transmit their genotypes directly to their offspring. Instead, they mate and contrib-

BOX 7.1 Haploid Selection

Equation 7.6 in the text gives us the allele frequency at any time in the future if we know the initial frequency $f_A(0)$ and the selection intensity. We can use that equation to find how long it takes for the frequency to reach a specific value, say f_A^*. To do that, we solve for $(1-s)^t$:

$$(1-s)^t = \frac{f_A(0)/f_A^* - f_A(0)}{f_a(0)}$$

Then we take the logarithm to obtain t:

$$t = \frac{1}{\log(1-s)} \log\left[\frac{f_A(0)/f_A^* - f_A(0)}{f_a(0)}\right]$$

If s is small, then $\log(1-s) \approx s$, so the time is inversely proportional to the selection coefficient. Therefore, if s is increased by some multiple of s, from s to Cs, then the time needed to reach the final frequency is reduced by the same factor, from t to t/C.

To illustrate this result, we can calculate that if $s = 0.01$, it takes roughly 277 generations to increase f_A from 0.2 to 0.8. If $s = 0.001$, then it takes roughly 2770 generations, and if $s = 0.1$, it takes roughly 28 generations.

ute only half their genes to each offspring. We will develop the basic ideas about selection by assuming that selection results from differences in the survival rate of individuals with different genotypes. That is the easiest type of selection to analyze, but the principles are the same for other kinds of selection, as well.

Viability selection is the kind of selection Darwin described as resulting from differences among individuals in their abilities to compete in the "struggle for existence." Darwin described selection based on differences in phenotype but we now know that differences in phenotype result in part from differences in genotype. In a few cases, phenotypic differences are attributable to different alleles at a single locus. For example, a major difference in coat color of beach mice (*Peromyscus polionotus*) that live on sandy islands in the Gulf Coast of Florida is determined by alleles at the melanocortin-1 receptor (*MC1R*) locus. This gene affects the production of melanin, which partly determines hair and skin color in mice, humans, and other mammals. In the beach mouse, there are two alleles, *R* and *C*. *RR* mice are dark in color, *CC* individuals are light, and *RC* individuals are intermediate (**Figure 7.3**). Coat color affects viability because it provides camouflage. Light-colored mice are more difficult to see on a sandy background than darker mice, and so they are eaten less often by visual predators such as hawks. Not surprisingly, the lighter-colored form is more prevalent on beaches, and the darker-colored form is more common in woods and grassy areas, where the substrate is darker.

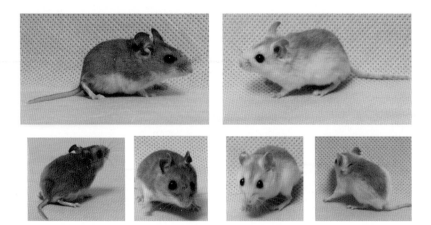

Figure 7.3 Photographs of light- and dark-colored forms of the beach mouse. (From Hoekstra et al., 2006.)

The islands where the lighter-colored mice live have been above water for only a few thousand years. The populations found on the sandy islands arose from mainland populations that are darker in color and lack the C allele. The C probably arose as a mutation after the mice had established populations on the islands. Then C increased in frequency because CR and CC mice had better chances of surviving to adulthood. This is one of many examples in which natural selection has led to an obvious evolutionary change in a relatively short time.

To understand just how selection causes allele frequencies to change, we need to quantify the effect of each genotype on viability, which is the probability of surviving from the zygote stage to the adult stage. For one locus with two alleles, there are three genotypes, and each may have its own viability: v_{AA}, v_{Aa} and v_{aa}. These viabilities are averages over large numbers of zygotes that have the same genotype at this locus. Each zygote will have different genotypes at other loci and will develop into an individual who experiences the environment in a unique way. But when averages are taken, any difference the genotype makes to survival will become apparent. For the beach mouse, v_{RC} is estimated to be between 80% and 90% of v_{CC}. That means that a slightly darker mouse living on a sandy background has a 10%–20% lower chance of surviving to adulthood than does a lighter-colored mouse.

To understand the effects of viability selection, assume that a population mates and produces offspring all at the same time. Among the parents, let the frequency of A be f_A, and assume the parents mate randomly to produce a large number of zygotes, N, that start the next generation. Because of random mating, we know that the genotype frequencies of the zygotes are the Hardy–Weinberg frequencies f_A^2, $2f_A f_a$, and f_a^2, so the numbers of zygotes with each of the three genotypes will be Nf_A^2, $2Nf_A f_a$, and Nf_a^2. **Box 7.2** presents a numerical example. The zygotes develop into juveniles,

---

BOX 7.2   One Generation of Viability Selection

Suppose we could study 100,000 newborn mice produced by a group of parents in which the frequency of C is 0.3, and suppose that the viabilities of the three genotypes are $v_{CC} = 0.5$, $v_{CR} = 0.4$, and $v_{RR} = 0.3$. If the newborns are exactly in Hardy–Weinberg proportions, there will be 9000 CC; 42,000 CR; and 49,000 RR.

The number of each genotype that survives to adulthood depends on the viabilities: $0.5 \times 9000 = 4500$ CC adults; $0.4 \times 42,000 = 16,800$ CR adults; and $0.3 \times 49,000 = 14,700$ RR adults. The genotype frequencies among the adults are found by dividing the number with each genotype by 36,000, the total number of survivors (4500 + 16,800 + 14,700):

$$f_{CC} = 4500/36,000 = 0.125; f_{CR} = 16,800/36,000$$
$$= 0.467; \text{ and } f_{RR} = 14,700/36,000 = 0.408$$

The frequency of C among the surviving adults is $f_C = f_{CC} + f_{CR}/2 = 0.125$ + 0.233 = 0.358. Random mating does not change the allele frequency, so we have determined that f_C has increased from 0.3 to 0.358 after one generation of viability selection.

We can see that only the relative viabilities and not their magnitudes determine the change in allele frequency. For example, suppose that all three viabilities are $^1/_{10}$ of the values given above: $v_{CC} = 0.05$, $v_{CR} = 0.04$, and $v_{RR} = 0.03$. In that case, fewer mice (3600) would reach adulthood (450 CC, 1680 CR, and 1470 RR). But the genotype frequencies in the adults will be the same as calculated above. The same would be true if all three viabilities were $^1/_{100}$ or some other fraction of the values first given.

---

which, if they survive, become adults. The numbers of adults with each of the three genotypes is the product of the initial number and the probability of survival. Box 7.2 computes the numbers of adults with the three genotypes for the viabilities estimated for mice on a light background. Now we have a group of adults who can mate randomly to create zygotes that form the next generation.

Each pair of parents produces on average more than two zygotes, sometimes vastly more, so the population size does not decrease. If the number of offspring for every pair of parents is the same on average, say r, then the average contribution of each genotype to the next generation is r multiplied by the viability: $w_{AA} = rv_{AA}$, $w_{Aa} = rv_{Aa}$ and $w_{aa} = rv_{aa}$. These w's correspond to the w's in the haploid model.

If we assume that the surviving adults mate randomly, the genotype frequencies in the zygotes in the next generation is determined by the allele frequencies among the adults in the previous generation. We compute that allele frequency as shown in Box 7.2. We can see that the difference in viabilities of the different genotypes result in a change in allele frequency.

BOX 7.3 Algebraic Calculation of Allele Frequency Changes

The change in allele frequency under arbitrary viabilities can be determined the same way as in Box 7.2. Suppose alleles A and a at a locus affect viability and the probabilities of surviving to adulthood are v_{AA}, v_{Aa}, and v_{aa}. If the allele frequencies among parents are f_A and f_a, and the parents mate randomly, the genotype frequencies among the offspring are given by the Hardy–Weinberg formulas. If there are N offspring, the numbers with each of the three genotypes are:

$$Nf_{AA} = Nf_A^2;\ Nf_{Aa} = 2Nf_A f_a;\ f_{aa} = Nf_a^2$$

The numbers of individuals with each genotype that survive to adulthood are:

$$Nv_{AA}f_A^2;\ 2Nv_{Aa}f_A f_a;\ Nv_{aa}f_a^2$$

The genotype frequencies among the adults are found by dividing each of these numbers by the total number of adults, which is:

$$N\left(v_{AA}f_A^2 + 2v_{Aa}f_A f_a + v_{aa}f_a^2\right) = N\bar{v}$$

It is convenient to denote the sum in parentheses by $\bar{v}$ because that is the average viability in the population. The genotype frequencies are then:

$$f'_{AA} = \frac{v_{AA}f_A^2}{\bar{v}};\ f'_{Aa} = \frac{2v_{Aa}f_A f_a}{\bar{v}};\ f'_{aa} = \frac{v_{aa}f_a^2}{\bar{v}}$$

where we have used a prime (′) to indicate the genotype frequencies after differences in viability have their effect. We canceled the N from the numerator and denominator of each fraction. We can see now why only the relative viabilities matter. If we multiply all three v's by the same constant, that constant will cancel from the numerator and denominator, leaving the genotype frequencies unchanged.

Random mating of the surviving adults does not change the allele frequency, so the frequency in the next generation is:

$$f'_A = \frac{v_{AA}f_A^2 + v_{Aa}f_A f_a}{\bar{v}}$$

The simple calculation in Box 7.2 shows that what we found for selection on haploids is true for selection on diploids as well: only the relative viabilities matter. The allele frequency in the next generation depends on the ratio of the viability of each genotype to the average viability, which is the denominator used to determine the allele frequencies. Genotypes that have a greater-than-average viability contribute more to the next generation than do genotypes that have a smaller-than-average viability.

In **Box 7.3**, we compute the allele frequency in the next generation for any set of viabilities. The algebra shows that in general, the ratios of the viabilities determine the change in allele frequencies. Because of that, we define a selection coefficient for each genotype in terms of the ratio of its

BOX 7.4 Special Cases of Selection

The last formula in Box 7.3, can be simplified and tell us how selection works in some special cases.

Additive selection: First assume that $v_{AA} = 1$, $v_{Aa} = 1 - s$ and $v_{aa} = 1 - 2s$. Each copy of the deleterious allele a reduces viability by s relative to the viability of AA individuals. We can simplify the formula for f'_A after substituting these viabilities. First:

$$\bar{v} = f_A^2 + 2(1-s)f_A f_a + (1-2s)f_a^2 = f_A^2 + 2f_A f_a + f_a^2 - 2s(f_A f_a + f_a^2) = 1 - 2sf_a$$

Then:

$$f'_A = \frac{f_A^2 + (1-s)f_A f_a}{1-2sf_a} = \frac{f_A^2 + f_A f_a - sf_A f_a}{1-2sf_a} = \frac{f_A - sf_A f_a}{1-2sf_a}$$

We can simplify further by computing the change in allele frequency in one generation:

$$\Delta f_A = f'_A - f_A = \frac{f_A - sf_A f_a}{1-2sf_a} - f_A = \frac{sf_A f_a}{1-2sf_a}$$

Selection coefficients are usually small: $s = 0.1$ is considered large. If s is small, then the denominator is approximately 1, and we get $\Delta f_A \approx sf_A f_a$.

This expression is convenient because it can be further approximated by the differential equation for allele frequency as a function of time,

$$\frac{df_A}{dt} = sf_A(1-f_A)$$

which can be solved.

We can carry out a similar analysis for two other cases that commonly arise.

Dominant advantageous allele: $w_{AA} = w_{Aa} = 1$ and $w_{aa} = 1 - s$.

$$\Delta f_A = \frac{sf_A f_a^2}{1-sf_a^2}$$

Recessive advantageous alleles: $w_{AA} = 1$ and $w_{Aa} = w_{aa} = 1 - s$.

$$\Delta f_A = \frac{sf_A^2 f_a}{1-s(2f_A f_a + f_a^2)}$$

viability to the largest viability. When AA individuals have the highest viability, the selection coefficients for Aa and aa are:

$$\frac{v_{Aa}}{v_{AA}} = 1 - s_{Aa}; \quad \frac{v_{aa}}{v_{AA}} = 1 - s_{aa} \tag{7.7}$$

In the example of the mice, $s_{Aa} = 0.1$–0.2. The selection coefficient provides a convenient way to characterize selection.

Several special cases of viability selection are useful to distinguish because the change in allele frequency can be easily expressed (**Box 7.4**). Alleles are *additive* in their effect on viability if $s_{aa} = 2s_{Aa}$. For alleles with a small

---

BOX 7.5 Genic Selection

For genic selection, each copy of a changes the viability by the same factor, $1 - s$: $w_{Aa} = w_{AA}(1-s)$ and $w_{aa} = w_{AA}(1-s)^2$. Substituting these expressions into the last equation in Box 7.3, we get:

$$f'_A = \frac{w_{AA}f_A^2 + (1-s)w_{AA}f_A f_a}{w_{AA}f_A^2 + (1-s)w_{AA}2f_A f_a + (1-s)^2 w_{AA}f_a^2} = \frac{f_A}{f_A + (1-s)f_a}$$

which is the same as Equation 7.6 with $t = 1$. Genic selection is equivalent to selection in haploids, because each copy of A makes the same contribution to viability whether it is in a heterozygous or a homozygous individual.

---

additive effect on viability (i.e., $s_{aa} < 0.1$), Box 7.4 shows that the change in allele frequency in one generation is approximately $s_{Aa}f_A f_a$. This is called an additive model of selection because each a reduces the viability by the same amount. **Box 7.5** also presents the results for the case with A dominant ($v_{AA} = v_{Aa}$) and recessive ($v_{Aa} = v_{aa}$) in its effect on viability.

To predict the time course of allele frequency in the population, we can repeat the calculation for one generation of viability selection for as many generations as we want, assuming the viabilities do not change. **Figure 7.4** shows the predictions for fifty generations of the mouse example. The results were obtained using a simple computer program that repeats what was done in Box 7.2 fifty more times.

In the special case of **genic selection**, a diploid population is equivalent to a haploid population. There is genic selection if each copy of a reduces viability by a factor $(1 - s)$, so that $v_{Aa}/v_{AA} = 1-s$ and $v_{aa}/v_{AA} = (1-s)^2$. Box 7.5 shows that the allele frequency changes when there is genic selection as given by Equation 7.6. If s is small, then $(1-s)^2 \approx 1 - 2s$, so the model of genic selection is almost the same as a model of additive selection (Box 7.4). Therefore, Equation 7.6 provides a good approximation to additive selection as well.

When an allele (say, A) is in low frequency, its change in frequency is determined by the ratio of the viability of the heterozygote to the viability of the other homozygote (aa), unless the allele is recessive in its effect on viability. If A is rare, the initial change in f_A will depend on v_{Aa}/v_{aa}. The reason is that when an allele is in low frequency, individuals homozygous for it are so rare that the viability of homozygotes makes little difference. Only when an allele becomes more common do the homozygotes become important. As a consequence, it is easy to determine whether an allele will increase in frequency when it is rare. All you have to do is ask whether individuals heterozygous for that allele have a higher viability than individuals homozygous for the other allele.

We can predict what will happen after a large number of generations just by knowing the selection coefficients. If the selection coefficients do not

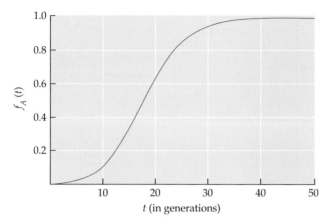

Figure 7.4 Allele frequencies for 50 generations with $s_{Aa} = 0.2$, $s_{aa} = 0.4$ (additive selection) and $f_A(0) = 0.01$.

change with time, there are only three possibilities. The first is **directional selection** in favor of one of the alleles. There is directional selection in favor of A if the aa individuals are less viable than AA individuals ($v_{aa} < v_{AA}$) and the viability of the Aa individuals is between those two, $v_{aa} \leq v_{Aa} \leq v_{AA}$. If there is directional selection in favor of A, f_A will increase every generation regardless of the initial frequency and eventually will approach 1. When f_A reaches 1, selection will have caused the fixation of A and the loss of a. A is the **advantageous allele** and a is the **deleterious allele**. In saying that, we always have to keep in mind that the terms advantageous and deleterious pertain only to a particular environment. In our mouse example, C is the advantageous allele in populations on the lighter substrate and R is the advantageous allele in populations on the darker substrate.

Although the long-term result of directional selection is always the same, the rate of change in allele frequency depends on the magnitudes of the selection coefficients. Not surprisingly, larger selection coefficients result in more rapid change. Roughly speaking, the time it takes for directional selection to result in a substantial change in allele frequency is $1/s_{aa}$. This statement is not precise but it gives us an approximate idea of how long it takes for a significant evolutionary change to occur. Viability differences of 1% result in substantial changes in allele frequency over hundreds of generations. Viability differences of 10% result in substantial changes over tens of generations. This point is illustrated in Figure 7.3. This result, which was obtained in the 1920s, played an important role in integrating Mendelian genetics into Darwin's theory of natural selection. It shows that even very weak natural selection, in which viabilities differ by a tenth of a percent or less, is strong enough to cause allele frequencies to change substantially in a few thousand generations. Although that is a long time for a scientist doing an experiment in a laboratory, it is short compared to

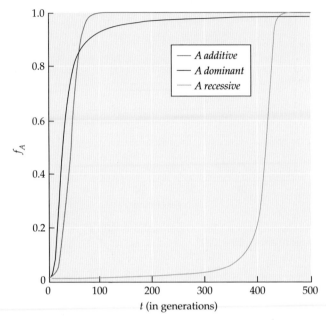

Figure 7.5 Illustration of the difference between selection on an advantageous allele that is dominant, recessive, or additive in its effect on viability. In all cases, $w_{AA} = 1$ and $w_{aa} = 0.8$.

the evolutionary history of most species. Natural selection acting on even minute differences in viability is more than up to the task of causing large-scale evolutionary changes.

If an allele is recessive in its effect on viability, selection is inefficient when it is in low frequency. For example, if $s_{Aa} = 0$, then a is recessive: Aa individuals have the same viability as AA individuals. In that case, selection has a difficult time removing a completely from the population because f_A changes only because of the lower viability of the aa individuals. If f_a is small, those individuals are in frequency f_a^2. We see that effect in Figure 7.4. In contrast, when A is recessive in its effect on viability (i.e., $s_{Aa} = s_{aa}$), selection has a difficult time increasing f_A when it is initially small. The advantage in viability is felt only by AA individuals and they are rare when f_A is small. In **Figure 7.5**, note the difference between A dominant and A recessive.

The second possibility is **heterozygote advantage**, which occurs when the heterozygote has the highest viability: $v_{Aa} > v_{AA}, v_{aa}$. In this case, we define the selection coefficients of the two homozygotes in terms of the viability of the heterozygotes:

$$\frac{v_{aa}}{v_{Aa}} = 1 - s_{aa}; \quad \frac{v_{AA}}{v_{Aa}} = 1 - s_{AA} \tag{7.8}$$

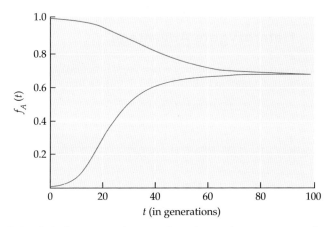

Figure 7.6 Allele frequency change when there is heterozygote advantage. w_{AA} = 0.9, w_{Aa} = 1, w_{aa} = 0.8, $f_A(0)$ = 0.01 or 0.99.

When we compute allele frequencies in future generations, we find that neither allele is eliminated by selection. Instead, f_A approaches the same value regardless of its initial value (**Figure 7.6**).

Heterozygote advantage maintains a **stable polymorphism**. We can see why this is so by considering what happens when one of the alleles is in low frequency. If A is low frequency ($f_A \ll 1$), then most of the population is aa and the rest is Aa; AA individuals are very rare. Because $v_{Aa} > v_{aa}$, heterozygous individuals have a higher viability and A will increase in frequency. For a similar reason, because $v_{Aa} > v_{AA}$, a will increase in frequency when it is rare. Natural selection cannot eliminate either allele because they both tend to increase when rare. Heterozygote advantage is a special case of **balancing selection**, selection that tends to maintain polymorphism by causing low frequency alleles to increase in frequency. Later, we will describe another kind of balancing selection that results from fertility differences in plants.

We can easily predict the equilibrium frequency from the selection coefficient:

$$\hat{f}_A = \frac{s_{aa}}{s_{AA} + s_{aa}} \tag{7.9}$$

where the carat ($\hat{\ }$) indicates the long-term equilibrium frequency. This result is derived in **Box 7.6**.

Examples of heterozygote advantage are rare but important. The most famous case is that of the S allele of the β-globin gene in human populations that have a high prevalence of malaria. In the 1950s, a very large study of survivorship provided direct evidence of heterozygote advantage and allows the selection coefficients to be estimated, s_{AA} = 0.075 and s_{aa} = 0.748. **Box 7.7** shows how these values are obtained.

BOX 7.6 Heterozygote Advantage

We can find the equilibrium frequency when there is heterozygote advantage by using the last equation in Box 7.3 and requiring that $f'_A = f_A$:

$$f_A \bar{w} = w_{AA} f_A^2 + w_{Aa} f_A f_a$$

Let $v_{Aa} = 1$, $v_{AA} = 1 - s_{AA}$ and $v_{aa} = 1 - s_{aa}$. Therefore:

$$\bar{v} = f_A^2(1 - s_{AA}) + f_{Aa} + f_a^2(1 - s_{aa}) = 1 - f_A^2 s_{AA} - f_a^2 s_{aa}$$

and

$$v_{AA} f_A^2 + v_{Aa} f_A f_a = (1 - s_{AA}) f_A^2 + f_A f_a = f_A - s_{AA} f_A^2$$

Substituting these into the top equation and simplifying gives:

$$s_{AA} f_A = s_{aa} f_a$$

from which Equation 7.9 follows. Note that this result does not assume that the selection coefficients are small.

Alleles at other loci in humans also seem to be affected by heterozygote advantage resulting from malarial infections. Several alleles at the *α-globin* locus that cause another kind of blood disorder, called thalassemia, also provide protection against malaria. *Thallasa* is the Greek word for sea, and thalassemia is the name given to this condition because it was common in countries around the Mediterranean Sea where malaria was prevalent until the last 200 years. The A^- allele of *G6PD*, discussed in Chapter 6 is another example of an allele that confers some protection against malaria. It is not known whether A^- is at equilibrium under balancing selection or whether it will continue to increase in frequency as long as malaria remains an important cause of infant and juvenile mortality in western Africa.

Although heterozygote advantage can maintain polymorphism, it may not be common. One of the ongoing debates within evolutionary biology is how important heterozygote advantage is for maintaining genetic polymorphism. Loci in the major histo-compatibility complex (MHC), which code for important components of the human immune system, provide the only other well-established examples of heterozygote advantage. The evidence for heterozygote advantage is indirect but compelling. Most polymorphic loci in humans have only two or three alleles. For example, there are three alleles at the ABO blood group locus. In contrast, several loci in the MHC region have hundreds of alleles each. This great diversity is an important reason that genetic matching of organ donors and recipients is so important. The more MHC alleles that are shared by donors and recipients, the weaker the rejection response by the recipient is and the more likely the transplant is to be successful. Although the biological cause of heterozygote advantage

BOX 7.7   Estimates of Selection Coefficients for the S Allele in a West African Population

In the early 1950s, it was proposed that the S allele of the β–globin locus conferred a survival advantage to heterozygous carriers even though individuals homozygous for S had low survival rates because they had sickle-cell anemia, a severe form of anemia that causes mortality in infants and children. In a study designed to test this hypothesis, a group of 30,923 adult were typed. The numbers of individuals with each genotype are shown in the first line of the table.

	AA	AS	SS
Numbers	25,374	5482	67
Frequencies	0.821	0.177	0.002
HW frequencies	0.826	0.165	0.008
HW expectation	25,542.4	5012.3	247.3
Newborns	33,040	6600	320
w	0.768	0.830	0.209

The results show a significant deviation from the Hardy–Weinberg frequencies ($\chi^2 = 167.6$, $P < 2 \times 10^{-38}$), because there are more AS adults and fewer AA and SS adults than expected.

To find the intensity of selection against AA and SS individuals, we need to compute the relative viabilities of the three genotypes. To do that, we need the total number of newborns (N), which we do not know, but which in fact does not matter because only the ratios of the viabilities affect s_{AA} and s_{aa}. We assume that S is at its equilibrium frequency ($p_A = 0.909$, $p_S = 0.091$) and that the newborns were in their HW frequencies, given in the third line of the table. Suppose that the individuals sampled were the survivors of 40,000 newborns. That tells us the viabilities, which are given in the last line of the table. The selection coefficients are $s_{AA} = 1 - v_{AA}/v_{Aa} = 0.075$ and $s_{aa} = 1 - v_{aa}/v_{Aa} = 0.748$. Assuming any other N will lead to the same estimates of the selection coefficients. Of course, if the observed frequency of A is not the equilibrium frequency, then we cannot estimate the selection coefficients this way.

is not known, it is likely that being heterozygous at MHC loci provides the best protection against pathogens that an individual has not been exposed to previously. If every heterozygous individual has higher viability, on average, than every homozygote, then every new mutation will tend to increase in frequency, resulting in a large number of different alleles being maintained at each locus.

Figure 7.7 Illustration of heterozygote disadvantage. $w_{AA} = 0.9$, $w_{Aa} = 0.8$, $w_{aa} = 1.0$, $f_A(0) = 0.65$ or 0.7

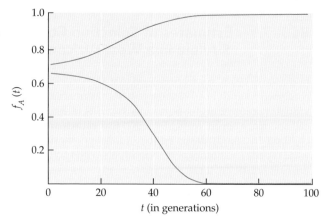

The third type of selection is **heterozygote disadvantage**, in which the heterozygote has the lowest viability, $v_{Aa} < v_{AA}, v_{aa}$. Heterozygote disadvantage results in the fixation of one of the alleles, but which allele is fixed depends on the initial allele frequency. As illustrated in Figure 7.6, the allele that is initially rare will be lost and the other will be fixed. We can deduce this by reversing the argument for heterozygote advantage. If A is initially rare, then most copies of A will be in heterozygotes, which have a lower average viability than the aa individuals. Therefore f_A will decrease. As **Figure 7.7** indicates, there is an intermediate frequency that separates the region for which A will be fixed from the region for which a will be fixed. Heterozygote disadvantage is one example of **disruptive selection**, which is selection that tends to remove low-frequency alleles. Disruptive selection can also result from genotype-specific fertility differences, as we will describe later in this chapter.

Good examples of heterozygote disadvantage are not common, or at least are not commonly detected. The reason is that it is difficult to distinguish heterozygote disadvantage that causes the loss of an allele from directional selection against that allele. A reduction in the fertility of heterozygotes is seen in individuals heterozygous for chromosomal rearrangements such as translocations. Translocations can cause gametic death because they result in the mispairing of chromosomes during meiosis. An interesting and important question in evolutionary biology is: If chromosomal rearrangements result in reduced fertility of heterozygotes, what causes populations of different species to differ by one or more chromosomal rearrangements?

Mutation–Selection Balance

A mutation occurs when there is an error in DNA replication during meiosis. The result is a change in a single nucleotide or the insertion or deletion of one or more nucleotides. Mutation creates a new allele at a locus, and we can ask what happens to it in subsequent generations. The answer depends on the mutation's average effect on viability. If a mutation does not affect the phenotype at all, it is neutral and subject only to genetic drift. If it results in lower viability, then it is deleterious, and selection will tend to reduce its frequency. We would expect each deleterious mutation to be lost because of selection against it. However, deleterious mutations of a single gene can arise repeatedly, so a balance may develop between the creation of deleterious alleles by mutation and their elimination by selection. We can use what we have learned about selection to determine what frequency of deleterious mutations we would expect to see if there is a **mutation–selection balance**.

Many genetic diseases in humans are caused by deleterious mutations of single genes. Such diseases are called **monogenic** or **Mendelian** diseases, to distinguish them from complex diseases, discussed in Chapter 6. An example of a monogenic disease is phenylketonuria (PKU), which is caused by a recessive mutation of the phenylalanine hydroxylase (*PAH*) gene. Individuals homozygous for this recessive allele are unable to metabolize phenylalanine. If they eat a normal diet from birth, they will develop severe mental retardation and die at an early age because of the accumulation of phenylalanine, which is a neurotoxin. Newborns in many countries are tested at birth for PKU. PKU babies are then put on a phenylalanine-free diet, and they develop normally. Until the dietary treatment was developed in the 1900s, PKU was a lethal condition. Although selection eliminates alleles causing PKU, new alleles are created by mutation, resulting in a incidence of about 1 in 10,000 in the United States.

Suppose there is a population in which A has a frequency of nearly 1 and assume that the probability that each copy of A mutates to a deleterious form a is μ per generation. Mutation increases f_a by μf_A, which is approximately μ, per generation. First assume that a is not completely recessive—that is, s_{Aa} is not zero. Because f_a is small, almost all copies of A are in heterozygotes, which have a frequency of $2f_A f_a$ if mating is random. If f_A is nearly 1, the frequency of heterozygotes is approximately $2f_a$. Because each heterozygote has a viability of $1-s_{Aa}$ compared to the viability of aa individuals, selection will reduce the frequency of heterozygotes from $2f_A$ to approximately $(1-s_{Aa})2f_A$ in one generation. The decrease in f_a due to selection is then $s_{Aa}f_a$: the factor of 2 disappears because each heterozygote carries only one a. The net change in f_a between two generations is the gain caused by mutation minus the loss caused by selection:

$$\Delta f_a = \mu - s_{Aa}f_a \tag{7.10}$$

A balance between mutation creating new deleterious alleles and selection removing them is attained when $\Delta f_A = 0$, which occurs at a frequency

$$\hat{f}_a = \frac{\mu}{s_{Aa}} \tag{7.11}$$

This result makes intuitive sense. If $s_{Aa} = 1$, which means that heterozygotes do not survive at all, then $\hat{f}_a = \mu$. That means that every copy of a has to be a mutation that has just occurred, because in previous generations, no Aa individuals survived to mate. If s_{Aa} is smaller than 1, the frequency of a is correspondingly larger. Note that the frequency is proportional to the mutation rate, μ. Selection against deleterious alleles that affect heterozygote viability is very effective. A typical mutation rate for functionally different alleles in the fruit fly *Drosophila melanogaster* is $\mu = 10^{-6}$. Even if s_{Aa} is only $0.01, \hat{f}_a = 10^{-4}$.

Deleterious recessive alleles will be more frequent because selection is less effective at removing them. If a is recessive and deleterious, then selection reduces its frequency only because of the reduced viability of aa homozygotes. The frequency of a is reduced from f_a^2 to approximately $f_a^2(1 - s_{aa})$, hence the reduction is $s_{aa} f_a^2$. Mutation from A to a increases f_a by μ per generation, assuming f_A is nearly 1. Therefore, the net change per generation is

$$\Delta f_a = \mu - s_a f_a^2 \tag{7.12}$$

and the balance is attained at a frequency

$$\hat{f}_a = \sqrt{\frac{\mu}{s_{aa}}} \tag{7.13}$$

The square root in this expression makes a big difference. For example, if a is lethal ($s_{aa} = 1$) and $\mu = 10^{-6}$, then $\hat{f}_a = 0.001$. If the genotypes are at Hardy–Weinberg frequencies, then the frequency of heterozygous carriers is $2\hat{f}_a (1 - \hat{f}_a) \approx 0.002$. In other words, 1 in 500 individuals in the population will carry a lethal allele at this locus. The frequency of PKU at birth in the United States is about $1/10,000$. Assuming genotypes are in Hardy–Weinberg proportions, that implies the frequency of the causative allele is $1/100$. That frequency is typical of lethal recessive genetic diseases.

ALLELIC HETEROGENEITY   For PKU and other Mendelian diseases, the classical and modern definitions of an allele are slightly different and can create some confusion. The difference is a consequence of how an allele is detected. Before DNA sequencing could be performed, an allele was defined by the phenotype associated with it. PKU was recognized as a Mendelian recessive condition because its appearance in families was consistent with the assumption that there was an allele that caused PKU in homozygous individuals. Once the locus responsible for the vast majority of PKU cases was identified as being caused by mutations in the *PAH* gene, it became possible to compare causative alleles in different families. In PKU and in other Mendelian diseases, many different defects result in the same phenotype. In the *PAH* gene, more than 500 different defects are known to cause PKU. Of the known alleles, roughly 63% are nonsynonymous changes, 13% are small deletions, 11% are changes in splice sites that lead to incorrect excision of introns, 7% are synonymous changes, 5% are changes to stop codons, and 1% are small insertions. This diversity tells us there is **allelic heterogeneity** at *PAH*. What appears to be one allele that causes a particular condition is actually a heterogeneous group of alleles, each of which has the same effect on phenotype.

The theory of mutation–selection balance described in the previous section applies when mutation consistently creates new alleles that replace ones lost due to selection. For PKU, the theory applies to the class of alleles that cause the disease but not to each individual allele in the class. A vast majority of disease-causing alleles arise only once by mutation, persist for some time, and then are lost. They are replaced not by identical mutations, but by different mutations that have the same effect on phenotype.

FERTILITY SELECTION   **Fertility selection** occurs when the number of offspring produced by a mating pair depends on the genotypes of the parents. An example of fertility selection is provided by the Rh blood group system in humans. Individuals have either the Rh factor (the D antigen) and are "Rh positive" (Rh⁺) or they lack it and are "Rh negative" (Rh⁻). The Rh system is, after the ABO system, the most important factor for determining the success of transfusions. Fertility selection occurs at the Rh system because an Rh⁻ mother may develop antibodies against the Rh factor of her Rh⁺ fetus, resulting in a potentially serious hemolytic disease called Rh disease. This is an example of fertility selection, because the viability of the fetus depends not on its own genotype, but on the genotypes of its parents.

We can illustrate how to analyze fertility selection by considering how selection affects the Rh system. The Rh factor is produced by a dominant allele, R, at the *RHD* locus on chromosome 1. Given the dominance, there are three types of families, shown in **Table 7.1**. Only when the mother is rr and the father carries at least one R is offspring survival reduced. Although R is dominant, $Rr \times rr$ families will suffer half the loss in fertility as $RR \times rr$ families, because only half of their offspring will be Rh⁺, while all of the offspring of an $RR \times rr$ family will be Rh⁺.

Unlike the case with viability selection, we cannot assume Hardy–Weinberg genotype frequencies among newborns. Therefore, we have to keep track of the genotype frequencies instead of the allele frequencies. The full analysis of this model is quite complicated, but by considering what happens when R or r is in low frequency, we can see that this model of selection is similar to a model with heterozygote disadvantage (Figure 7.7). If R is in very low frequency, then an Rr male is almost certain to marry an rr female.

TABLE 7.1 Genotypes of families of the *Rh* system

Father	Mother	Frequency	Offspring viability
RR	rr	$f_{RR}f_{rr}$	$1-2s$
Rr	rr	$f_{Rr}f_{rr}$	$1-s$
RR, Rr	RR, Rr	$(1-f_{rr})^2$	1
rr	RR, Rr, rr	f_{rr}	1

The result will be that he has fewer children than an *rr* male. Therefore, *R* will decrease in frequency. On the other hand, if *r* is in low frequency, most of the marriages of *rr* females will be with *RR* or *Rr* males, so these females will have fewer children than *Rr* or *RR* females. The result is that when *r* is rare, it will decrease in frequency. Selection against *r* when it is rare is relatively weak, because it is felt only in *rr* homozygous females. The fact that the *Rh* gene is polymorphic in humans suggests that other factors are responsible for maintaining that polymorphism. Maternal–fetal incompatibility alone would tend to eliminate one or the other allele.

Another type of fertility selection is important in plants. In many plants, self-fertilization is prevented because a locus, called the *S* locus, carries alleles that result in the infertility of certain mating pairs. There are several types of self-incompatibility systems. A common one is **gametic self-incompatibility**. Many plant families, including the Solanaceae (nightshades) and Rosaceae (roses) have numerous species with gametophytic self-incompatibility. To describe it, we denote the different alleles at the *S* locus by S_1, S_2, etc. If there is gametic self-incompatibility, pollen carrying a particular *S* allele, S_k, cannot fertilize a plant with the genotype S_iS_j ($i \neq j$) if $k = i$ or if $k = j$. For example, pollen carrying S_1 cannot fertilize plants with genotypes S_1S_2, S_1S_3, etc. Fertilization cannot occur because factors in the stigma prevent the pollen tube from growing. This mechanism prevents self-fertilization, but it also prevents cross-fertilization by some other plants in the population. One consequence is that no plants can be homozygous for an *S* allele. Another consequence is that there have to be at least three *S* alleles. Typically there are ten or more.

It is easy to see that self-incompatibility is similar to heterozygote advantage. If a mutation to a new *S* allele occurs, then pollen that carries that allele will be able to fertilize every other plant in the population. On average, the plant heterozygous for the new allele will have more offspring than all the other plants, whose fertility is reduced somewhat because some of their pollen lands on plants with incompatible *S* genotypes. As a consequence, low-frequency alleles tend to increase in frequency. If all *S* alleles are equivalent to one another, then their frequencies will be equal when the population reaches equilibrium.

References

*Avent N. D. and Reid M. E., 2000. The Rh blood group system: a review. *Blood* 95: 375–387

Bollback J. P. and Huelsenbeck J. P., 2007. Clonal Interference Is Alleviated by High Mutation Rates in Large Populations. *Molecular Biology and Evolution* 24: 1397–1406.

Cavalli-Sforza L. L. and Bodman W. F., 1971. *The Genetics of Human Populations*. San Francisco: W. H. Freeman.

*Haring V., Gray J. E., McClude B. A., et al., 1990. Self-Incompatibility: A self-Recognition System in Plants. *Science* 250: 937–941.

Hoekstra H. E., Hirschmann R. J., Bundey R. A., et al., 2006. A single amino acid mutation contributes to adaptive beach mouse color pattern. *Science*, 313: 101–104. http://www.sciencemag.org/content/313/5783/101.full

Scriver C. R., 2007. The PAH gene, phenylketonuria, and a paradigm shift. *Human Mutation* 28: 831–845.

*Vogel F. and Motulsky A. G, 1996. *Human Genetics: Problems and Approaches*, Third Edition. New York: Springer-Verlag.

*Recommended reading

EXERCISES

7.1 Suppose that a new allele *A* is created by mutation in a haploid species and that *A* results in a 1% higher growth rate per unit time.

   a. How long will it take for *A* to increase from 10% to 90% in frequency?

   b. How long will it take if the growth rate is 0.1% higher?

7.2 Suppose that a mutant allele *A* arose in a haploid population at an unknown time in the past. You know that its frequency today is 0.9. How many generations in the past did the mutation occur if the population size is 10,000? (Ignore the effects of genetic drift.) What about a population size of 100,000?

7.3 Suppose that *A* in Exercise 7.1 has a 5% lower growth rate on average. How long will it take for *A* to decrease from a frequency of 10% to a frequency of 1%?

7.4 Cystic fibrosis (CF) is a Mendelian recessive disease of humans caused by defects in ion transport (OMIM 602421[1]). Until the 1950s, when antibiotics were first used to treat CF patients, most newborns with CF died at an early age. Yet CF is relatively common in Caucasians, with a frequency at birth of 1/2500, which implies that the frequency of CF-causing mutations is about 0.02—a surprisingly high frequency of an allele that is lethal to homozygotes. There is no agreement on the reason for this high frequency.

   a. Suppose that an allele that causes CF is maintained by mutation–selection balance. What would be the mutation rate necessary for that allele to have a frequency of 0.02?

   b. Suppose that an allele that causes CF is maintained by heterozygote advantage. In order for the equilibrium frequency to be 0.02,

[1]OMIM = Online Mendelian Inheritance in Man (http://www.ncbi.nlm.nih.gov/omim), a website that provides detailed and authoritative information about genetic variants and genetic diseases in humans. The OMIM number refers to the specific entry.

what would the difference between the viabilities of the homozygote and the heterozygote have to be?

7.5 The frequency of the A^- allele of *G6PD* in western African populations is about 11%. If A^- is at its equilibrium frequency, what is the selection coefficient against normal (*BB*) homozygotes if the individuals homozygous for A^- have a 50% chance of surviving to adulthood in western Africa? (Ignore the presence of other alleles.)

7.6 A locus with two alleles, *B* and *b*, affects the viability of seeds of a plant population. One-fifth of the *BB* seeds germinate and produce adult plants; ⅙ of the *Bb* seeds germinate and produce adult plants; and ¹⁄₁₀ of the *bb* seeds germinate and produce adult plants. Fertility does not depend on the genotype at the *B/b* locus. If the frequency of *B* is ¼ in one generation and the genotypes in that population of seeds are in Hardy–Weinberg equilibrium, what will be the frequency of *B* in the seeds in the next generation?

7.7 *A* is the normal allele at the *β-globin* locus, but in a malarial region of western Africa, the *S* allele of this locus is present at a frequency of 0.2. *SS* individuals have sickle-cell anemia and have only a 10% chance of surviving to reproductive age relative to heterozygous individuals with the *AS* genotype. Normal individuals with the *AA* genotype at this locus have an 85% chance of surviving to reproductive age, relative to *AS* individuals. Assume that at this locus, the genotypes of newborns are in Hardy–Weinberg proportions.

a. If the relative fitness of *AS* individuals is 1, what is the average viability in this population?

b. What are the genotype frequencies among individuals of reproductive age?

7.8 Suppose that a new sand-covered island is created in the Gulf Coast of Florida and that the island is colonized by a population of *Peromyscus polionotus* that is fixed for the dark-colored allele of *MC1R*. If the population grows to 10,000 individuals and then an individual heterozygous for the light-colored allele arrives on the island and mates with one of the residents, how many generations will it take for the light-colored allele to reach a frequency of 99%? (Assume genic selection in favor of the light-colored allele and ignore the effects of genetic drift.)

7.9 Suppose you are concerned with the fertility differences caused by the genotype at a locus with two alleles, *A* and *a*. Suppose that all mating pairs produce the same number of offspring except for the *aa* × *aa* matings, which produce only half as many as the others. All the genotypes have the same viability.

a. If initially the frequencies of the three genotypes are ¼ *AA*, ½ *Aa*, ¼ *aa*, what will the genotype frequencies be after one generation of random mating?

b. Are the genotype frequencies in the newborns in their Hardy–Weinberg proportions?

7.10 Reciprocal translocations occur when there is an exchange of genetic material between nonhomologous chromosomes. Often reciprocal translations have no effect on phenotype, because there is a full complement of genes. But they reduce fertility of heterozygotes by a factor of ½ because half of the gametes produced are aneuploid. Assume that a population carries a reciprocal translocation that has no effect on viability. Let *A* be the translocation. Assume that initially, $f_{AA} = 0.01, f_{Aa} = 0.18$, and $f_{aa} = 0.81$. That is, *A* is initially in HWE with frequency $f_A = 0.1$.

a. What are the fertilities of each possible mating pair?

b. Explain why the effect of a translocation results in disruptive selection.

7.11 Suppose the *R* allele of the Rh system to be recessive in its effects, instead of dominant. That is, only *RR* individuals are Rh⁺; *Rr* and *rr* individuals are Rh⁻.

a. Fill in a table corresponding to Table 7.1 in the text that lists the fitness loss in all types of families. (Recall that incompatibility occurs when an Rh⁻ mother carries an Rh⁺ fetus.)

b. Would this case also result in disruptive selection?

7.12 Suppose a very large plant population has five *S* alleles in equal frequency, 0.2.

a. What are the genotype frequencies in this population if there is random mating? (Hint: there are no plants homozygous for an *S* allele.)

b. Assume that mutation creates a sixth *S* allele. How many more offspring, on average, will the mutant plant have than any other plant in the population?

7.13 The average viability depends on the allele frequencies:

$$\bar{v} = f_A^2 v_{AA} + 2 f_A (1 - f_A) v_{Aa} + (1 - f_A)^2 v_{aa}$$

Draw graphs of $\bar{v}$ as a function of f_A for these three cases: $v_{AA} = 0.5$, $v_{Aa} = 0.4$, $v_{aa} = 0.3$; $v_{AA} = 0.4$, $v_{Aa} = 0.5$, $v_{aa} = 0.3$; $v_{AA} = 0.5$, $v_{Aa} = 0.3$, $v_{aa} = 0.4$.

8 *Selection in a Finite Population*

IN CHAPTER 7, we described how natural selection changes allele frequencies. By focusing on selection alone, we could describe in a simple way how selection changes allele frequencies, either driving one allele to fixation or maintaining two alleles in a population. In this chapter, we will consider the combined effects of selection and drift. We will see that sometimes genetic drift can be ignored and the theory in Chapter 7 applied, but at other times, particularly when one allele is in low frequency, genetic drift is important. By properly accounting for genetic drift and selection together, we can make useful predictions about rates of evolution at the level of DNA sequences. We can also predict patterns of genetic variation in a genomic region that indicate selection has acted recently. Although our presentation will initially be somewhat abstract, we will later describe practical applications of the theory.

Fixation Probabilities of New Mutations

We start by reviewing the results for neutral alleles. In Chapter 2, we showed that the probability that a new neutral mutation is ultimately fixed is $1/(2N)$, where N is the population size. Most neutral mutations are lost, but some persist. We can take a closer look at what happens to mutations by using a computer simulation. The simulation program assumes that there is random mating each generation of a population containing N diploid individuals. The program starts with an allele in frequency $1/(2N)$ and imitates the randomness of genetic drift in each generation until the allele is either fixed or lost. **Box 8.1** presents an outline of the simulation program used.

BOX 8.1 Simulating Trajectories

It is often useful to do computer simulations of population genetic processes in order to better visualize what happens. Many of the figures in this chapter are based on results from a simulation program that models selection and genetic drift in a population of N diploid individuals. The simulations assume the Wright–Fisher model described in Chapter 2. There are two alleles, A and a in a population of N diploid individuals. At the start of a generation, the frequency of A is f_A. The allele frequency is then changed to f'_A, given by the last equation in Box 7.3, because of genotypic differences in viability:

$$f'_A = f_A \frac{v_{AA}f_A + v_{Aa}(1 - f_A)}{v_{AA}f_A^2 + 2v_{Aa}f_A(1 - f_A) + v_{aa}(1 - f_A)^2}$$

Then genetic drift is simulated by having the computer choose $2N$ gametes randomly, each of which has a probability f'_A of being an A and a probability of $1 - f'_A$ of being an a. This is done for each gamete by generating a random number, x, that has a uniform distribution between 0 and 1. If $x < f'_A$, the gamete has an A and if $x \geq f'_A$ the gamete has an a. Then the program counts the number of A gametes to determine f_A in the next generation.

Each replicate simulation generates the trajectory of an allele. Then the selection and genetic drift steps are alternated until f_A reaches 0 or 1. After that, a new replicate is started and the process continues until a specified number of trajectories has been generated.

Running the simulation program for a new mutation produces a series of frequencies at times $t = 0, 1, 2 \ldots : f_A(0), f_A(1), f_A(2)\ldots$, until a time is reached when $f_A(t) = 0$ (loss) or $f_A(t) = 1$ (fixation). This series is the **trajectory** of that mutant and can be plotted as shown in **Figure 8.1**. We know the starting point, $f_A(0) = 1/(2N)$, because we assume the allele is a new mutant. The trajectory will be very short if the mutant is lost in the first few generations and much longer if the mutant continues segregating for a longer time. The

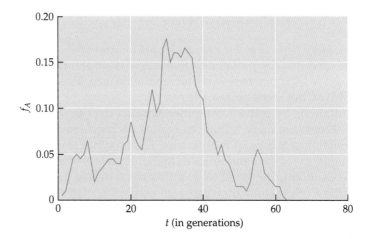

Figure 8.1 Example of a trajectory of a neutral allele. $N = 100$.

(A)

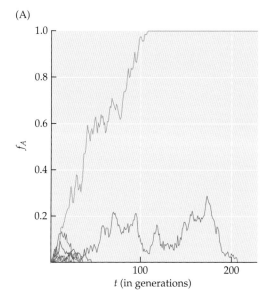

f_A

t (in generations)

Figure 8.2 (A) Fifty replicate trajectories of neutral alleles. $N = 100$. (B) Lower left corner of Figure 8.2A.

(B)

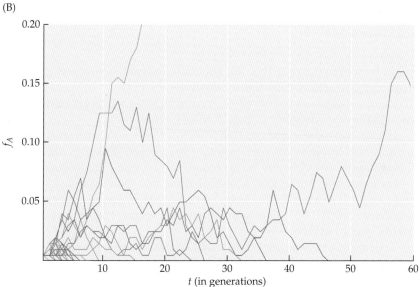

f_A

t (in generations)

trajectory for a single mutant is unpredictable, but the average properties of trajectories of a large number of mutants are quite predictable. A typical trajectory is one that goes to 0 within the first few generations. Figure 8.1 shows a longer trajectory in which A is lost after 61 generations.

For neutral alleles, our simulation program generates a wide range of trajectories (**Figure 8.2A**). **Figure 8.2B** shows more clearly the tangle of trajectories in the first few generations. One of the trajectories in Figure 8.2A is of a neutral allele that happens to go to fixation. That mutant steadily increases in frequency until it is fixed. If you saw only that trajectory, you

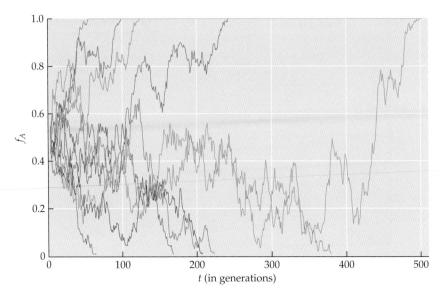

Figure 8.3 Fifty replicate trajectories of neutral alleles in a population of size $N = 100$ and $f_A(0) = 0.5$.

might be tempted to say that the allele was advantageous and driven to fixation by selection. This figure illustrates that it may be quite difficult to decide whether or not an allele is neutral, even when we have perfect knowledge of its frequency, since it arose by mutation.

If a neutral allele goes to fixation, it will do so within roughly $4N$ generations. Intuition may tell you that because an allele has no effect on survival and reproduction, it can remain at intermediate frequencies for a very long time, but in fact, a neutral allele that reaches an intermediate frequency will not remain there; it will continue to increase or decrease and eventually reach 0 or 1. **Figure 8.3** illustrates several trajectories of alleles that all started at frequency 0.5.

The fate of a neutral allele is like that of a gambler in a fair game, that is, a game in which each player has an equal chance of winning. Suppose one player starts with n chips and the other starts with m chips. In each round, they play for one chip and each player has a 50% chance of winning. If they agree to stop playing when one loses all the chips, then the probability that the first player will finish with all his chips is $n/(n + m)$. That is, each player's chance of winning depends on that player's initial fraction of the total number of chips. This result tells us what is well known to gamblers: the person with the largest initial stake has the better chance of ultimately winning all the chips. This is known as the "gambler's ruin" paradox. The paradox is that what seems to be a fair game is not really fair in the long run. In population genetics, the allele that is initially in higher frequency is the one more likely to be fixed, even though it has no selective advantage.

Selection favoring a new mutation increases the probability of fixation, but not as much as you might think. When an advantageous allele is in low frequency, it can still be lost because the few individuals who carry it happen not to survive and reproduce. The mathematical theory of selection and drift is beyond the scope of this book, but the main result is important and can be appreciated without understanding how it is derived. For a mutant allele A subject to additive selection with selection coefficient s, the probability that the mutant is ultimately fixed, u, is

$$u(s,N) = \frac{1-e^{-2s}}{1-e^{-4Ns}} \qquad (8.1)$$

provided that N is large and s is small in absolute value. This formula is valid whether A is advantageous ($s > 0$) or deleterious ($s < 0$).

This formula is graphed in **Figure 8.4A** for $N = 100$. For a given N, we can distinguish three ranges of selection intensity. We already know that $u(0,N) = 1/(2N)$. As s becomes large, u becomes approximately $2s$. In the graph, you can see that as s increases, u becomes a line with slope 2. With $N = 100$, that is a valid approximation when $s > 0.005$. If $u \approx 2s$, we say that A is *strongly advantageous*. For strongly advantageous alleles, the fixation probability does not depend on the population size. Roughly speaking, if $2Ns > 1$, alleles are strongly advantageous.

If selection is weaker, $-1 < 2Ns < 1$, then the probability that A will become fixed is close to that for a neutral allele. If $s > 0$, u is slightly larger than $1/(2N)$ and if $s < 0$, it is slightly smaller (**Figure 8.4B**). We characterize this range of selection coefficients as *nearly neutral*. For nearly neutral alleles, selection makes some difference, but population size is also important. For *strongly deleterious* alleles, $2Ns < -1$; the fixation probability is very small and decreases to 0 rapidly as s becomes more negative (Figure 8.4C).

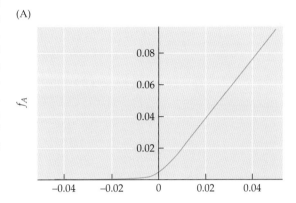

(A)

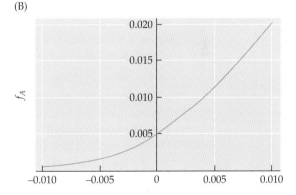

(B)

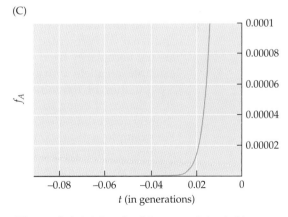

(C)

Figure 8.4 (A) Graph of Equation 8.1, $u(s,N)$ vs. s for $N = 100$. (B) Detail of Figure 8.4A for $-1/(2N) \leq s \leq 1/(2N)$ with $N = 100$. (C) Detail of Figure 8.4A for $-s \gg 1/(2N)$.

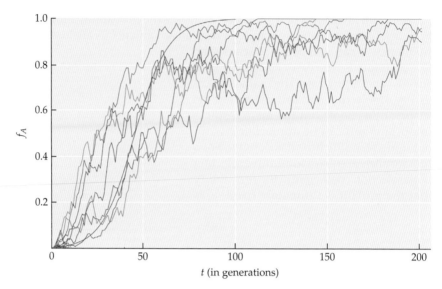

Figure 8.5 Fifty replicate trajectories for a strongly advantageous allele subject to additive selection with $s = 0.1$ and $N = 100$.

To summarize, for a population of a given size, the fixation probability of a new mutation falls in one of three ranges: strongly deleterious ($2Ns \ll -1$, $u \approx 0$), nearly neutral [$-1 < 2Ns < 1$, $u \approx 1/(2N)$], or strongly advantageous ($2Ns \gg 1$, $u \approx 2s$). These three selective regimes are only approximate. Figure 8.4A shows that $u(s, N)$ is a smooth function of s. There are no boundaries separating these ranges. Still, this provides a simple way to characterize what happens to new mutations and will be useful later when considering rates of molecular evolution.

Two implications of Equation 8.1 seem to violate both our intuition and what was presented in Chapter 7. First, strongly advantageous alleles are not necessarily fixed. They have a substantial chance of being lost. For example, an allele with additive effect and selection coefficient $s_{Aa} = 0.1$ in a population of 100 individuals is strongly advantageous but has only a 20% chance of being fixed. Second, in spite of the selection against them, slightly deleterious alleles have a small but non-zero chance of being fixed. For example, a disadvantageous mutant with additive effect and selection coefficient -0.001 has a 0.4% chance of being fixed. We see, then, that genetic drift can work against selection and result in a population that is not as well adapted as it could be. A population will lack some of the advantageous alleles that arose by mutation but did not go to fixation, and will be fixed for some slightly deleterious alleles that selection could not eliminate.

In thinking about these results, remember that whether an allele is strongly selected or nearly neutral depends on the population size (N) as well as on the selection coefficient (s). An allele that is strongly advantageous

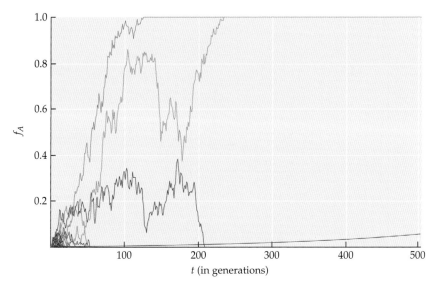

Figure 8.6 Fifty replicate trajectories of a weakly advantageous allele subject to additive selection. $s = 0.01$, $N = 100$.

in a large population may be nearly neutral in a small one. And an allele that is strongly deleterious and has essentially no chance of being fixed in a large population may become fixed in a small population by virtue of being nearly neutral.

The trajectories of selected alleles are also revealing. We used the simulation program described in Box 8.1 to generate sets of trajectories for alleles in the different selection regimes. **Figure 8.5** shows fifty trajectories for a strongly advantageous allele in a population of $N = 100$ individuals ($s_{Aa} = 0.1$ and $s_{aa} = 0.2$). Consistent with Equation 8.1, most mutants are lost. For the alleles that go to fixation, there is some regularity to the trajectories. They are similar to the trajectory predicted by Equation 7.6, which assumes that selection is acting alone (shown as the smooth curve in Figure 8.5). When an allele is strongly advantageous, it roughly follows the prediction based on selection alone once the frequency begins to increase. Genetic drift has some effect when the advantageous allele is rare, but selection is the predominant force once the allele becomes more common.

The picture is different for slightly advantageous alleles, as shown in **Figure 8.6**. Alleles that go to fixation do so at different times and follow trajectories that are quite different from what is predicted by the deterministic theory. The smooth line indicating the deterministic theory is just starting to increase after 500 generations, yet the mutants that go to fixation have done so by then.

For slightly deleterious alleles, a few alleles go to fixation, but most do not (**Figure 8.7**). For strongly deleterious alleles, the trajectories are all the same. An allele arises and is quickly lost (**Figure 8.8**).

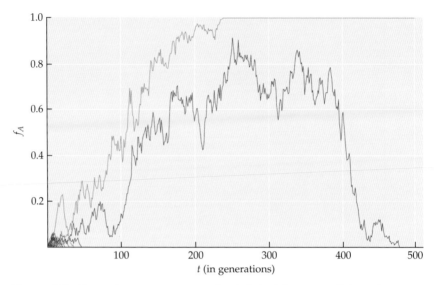

Figure 8.7 Fifty replicate trajectories of a slightly deleterious allele with an additive effect. $s = -0.005$, $N = 100$.

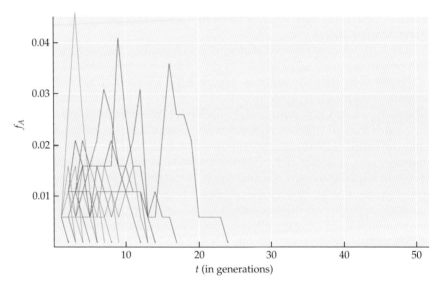

Figure 8.8 Fifty replicate trajectories of a strongly deleterious allele with an additive effect. $s = -0.1$, $N = 100$.

Rates of Substitution of Selected Alleles

In Chapter 2, we showed that the rate of substitution of neutral alleles is equal to the mutation rate, $r(0, N) = \mu$, where we have added the 0 to emphasize that this is the substitution rate for neutral alleles ($s = 0$) and the N to indicate the population size. This result follows from the facts that (1) the probability of fixation of a neutral allele is $u = 1/(2N)$ and (2) the number of alleles that appear each generation is $2N\mu$. Multiplying gives the number of new neutral alleles that appear and that also ultimately go to fixation each generation. We use the same logic to tell us the rate of substitution of new mutants that have a selection coefficient s. The number of mutations that appear each generation is $2N\mu$, and the probability of fixation is $u(s, N)$. Therefore:

$$r(s,N) = 2N\mu u(s,N) \qquad (8.2)$$

Equation 8.1 tells us that $u(s, N) > 1/(2N)$ if $s > 0$ and $u(s, N) < 1/(2N)$ if $s < 0$. Thus, advantageous alleles will be substituted at a higher rate than neutral alleles in the same population, while disadvantageous alleles will be substituted at a lower rate. Nearly neutral alleles will be substituted at almost the neutral rate, while strongly advantageous alleles will be substituted at a rate $r(s, N) = 4N\mu s$. The rate of substitution of strongly deleterious alleles is essentially 0.

You have already seen in Chapter 2 how to use the substitution rate for neutral alleles to estimate the time of separation of two species, using the formula

$$r = \frac{d}{2LT} \qquad (8.3)$$

where d is the number of differences in nucleotide sequence in a sequence of length L and T is the time separating the two species. Complexities arise when we compare species that diverged a long time in the past, because more than one substitution can occur at each site. **Box 8.2** presents one way to correct for multiple substitutions. For our purposes, the approximate formula, Equation 8.3, will be sufficient.

We can apply the same logic to nucleotide positions that might be subject to natural selection by comparing their substitutions rates with those for neutral positions. The first step is to analyze sites that we think are neutral, because for neutral sites, the substitution rate equals the mutation rate. For instance, **fourfold degenerate sites** in exons are considered to be neutral, because at these sites, a mutation does not change the amino acid coded for. As an example, the third position of the codon for proline is fourfold degenerate because each of the four codons—CCU, CCC, CCA and CCG—

BOX 8.2 Accounting for Multiple Substitutions

Equation 8.3 in the text relates the substitution rate (r) to the number of differences (d) found when comparing two DNA sequences. This formula is correct if every substitution that occurred during the time separating the two species ($2T$) is actually detected. If more than one substitution has occurred at a site, then d will not indicate the number of substitutions. It is possible that a second substitution undoes the effect of the first. For example, the first substitution might result in a C replacing a G, and then the second substitution replaces the G by a C. We would think that there has been no substitution at that site when in fact two have occurred. Or the second substitution could replace the G by a T. We would think that there has been one substitution, but in fact there have been two. The actual number of substitutions is at least as large as the number of differences counted, and possibly larger. The problem is to determine how many substitutions have actually occurred when the only information available is the number of differences in sequence.

A simple way to correct for multiple substitutions is to use the Jukes–Cantor model, which assumes that when a substitution occurs, each of the three other nucleotides is equally likely to be substituted. On this premise, the second substitution has a $\frac{1}{3}$ probability of undoing the first. For example, if the first substitution is a G for a C, then under the Jukes–Cantor model, the second substitution has a $\frac{1}{3}$ probability of being a C and a $\frac{2}{3}$ probability of being an A or a C. That is, we would have a $\frac{1}{3}$ chance of seeing no change and a $\frac{2}{3}$ chance of seeing one change when two substitutions have occurred.

The assumption leads to a formula that relates the actual number of substitutions that have occurred, K, to the number of differences detected, d:

$$K = -\tfrac{3}{4}\ln\left(1 - \tfrac{4}{3}\tfrac{d}{L}\right)$$

If d/L is much less than 1, then $\ln[1 - 4d/(3L)] \approx -4d/(3L)$ and we see that $K \approx d/L$, as is assumed in Equation 8.3. If d/L is larger, then K is larger than d/L.

This Jukes–Cantor model is one of a large number of models that have been developed to correct for multiple substitutions. More complicated models take into account other factors, such as the tendency for mutations to cause **transitions**—purine to purine (A to G, G to A) or pyrimidine to pyrimidine (C to T, T to C)—more frequently than **transversions**, purine to pyrimidine or pyrimidine to purine.

codes for proline. The "universal" genetic code, shown in **Table 8.1**, shows that the third position of codons for several other amino acids—leucine, valine, serine, threonine, alanine, and arginine—are also fourfold degenerate. It is reasonable to assume that mutations at a fourfold degenerate site

TABLE 8.1 Universal genetic code

Ala/A	GCU, GCC, GCA, GCG	Leu/L	UUA, UUG, CUU, CUC, CUA, CUG
Arg/R	CGU, CGC, CGA, CGG, AGA, AGG	Lys/K	AAA, AAG
Asn/N	AAU, AAC	Met/M	AUG
Asp/D	GAU, GAC	Phe/F	UUU, UUC
Cys/C	UGU, UGC	Pro/P	CCU, CCC, CCA, CCG
Gln/Q	CAA, CAG	Ser/S	UCU, UCC, UCA, UCG, AGU, AGC
Glu/E	GAA, GAG	Thr/T	ACU, ACC, ACA, ACG
Gly/G	GGU, GGC, GGA, GGG	Trp/W	UGG
His/H	CAU, CAC	Tyr/Y	UAU, UAC
Ile/I	AUU, AUC, AUA	Val/V	GUU, GUC, GUA, GUG
START	AUG	STOP	UAA, UGA, UAG

are neutral because they cause no change in the amino acid sequence of the protein coded for. Such a mutation is called **synonymous**. Therefore, by estimating substitution rates for fourfold degenerate sites, we can estimate the mutation rate at those sites.

As an example, consider the *β-globin* gene, There are 78 fourfold degenerate sites in the coding sequence ($L = 78$). When sequences from mice and humans are compared at those sites, 30 of them are found to differ ($d = 30$). The fossil record indicates that the most recent common ancestor of humans and rodents lived about 80 million years ago ($T = 8 \times 10^7$). From Equation 8.3, $r = 2.42 \times 10^{-9}$ per site per year. This is an estimate of the mutation rate for that gene, and is very close to the currently accepted average mutation rate for mammals.

All mutations that result in a change in amino acid are called **nonsynonymous**. In the universal code, all mutations in the second codon position are nonsynonymous. A simple way to estimate the rate of nonsynonymous substitutions is to count the number of differences in second-base codon positions.

Considering only fourfold degenerate sites and second codon positions is simple, but does not use all of the data. Methods for estimating synonymous and nonsynonymous substitution rates have to take into account both the idiosyncrasies of the genetic code and multiple substitutions. **Box 8.3** discusses two of the more important problems. Methods for estimating rates of synonymous and nonsynonymous substitutions are well developed, and freely available computer programs are available to implement them. **Table 8.2** presents some typical values for synonymous and nonsynonymous substitution rates for different protein-coding genes. These estimates are consistent with one another and support the assumption that synonymous

---

BOX 8.3 Computing Synonymous and Nonsynonymous Rates

Computing the fractions of synonymous and nonsynonymous differences between two aligned DNA sequences is not completely straightforward, because of the idiosyncrasies of the genetic code and the problem of allowing for more than one substitution. The underlying idea is simple. The fraction of synonymous differences is the number of synonymous differences divided by the total number of sites at which synonymous changes can occur. Complications arise because synonymous differences are not all the same. For example, suppose that in two sequences, the codon CTT (leucine) is aligned to CTT. There are no substitutions, but how many of the substitutions that did not occur are synonymous and how many are nonsynonymous? For this case, the answer is easy. Any substitutions in the first and second codon position are nonsynonymous and any substitution in the third codon position is synonymous, so we would say that there are two positions at which no nonsynymous substitution occurred and one position at which no synonymous substitution occurred. What about CAT (histadine) aligned to CAT? As with CTT, any substitution in the first or second position is nonsynonymous, but now two of the three substitutions in the third position (to an A or G) are also nonsynonymous. Only the substitution of C for T is synonymous. We would say that 2 $\frac{2}{3}$ of the potential nonsynonymous substitutions did not occur and only $\frac{1}{3}$ of a synonymous substitution did not occur.

Another problem arises when aligned codons differ by two or three nucleotides and the numbers of synonymous and nonsynonymous changes depends on the order in which those changes occur. For example, suppose that codon CAT (histadine) is aligned to CTG (leucine). If the intermediate codon is CTT (leucine) then the first change is nonsynonymous and the second is synonymous. If instead the intermediate codon is CAG (glutamine) then both substitutions are nonsynonymous.

---

mutations really are neutral. If selection were important, we would expect to see differences in the rate of synonymous substitutions among genes, but we don't. In contrast, rates of nonsynonymous substitution in protein-coding genes vary enormously. Some, including interleukin I, have rates that are not much lower than the synonymous rates. Others, including histones, have rates that are orders of magnitude lower.

Because the nonsynonymous rates in Table 8.2 are all lower than the rate of synonymous substitution, we can conclude that natural selection prevented some nonsynonymous substitutions from occurring. There is no reason to suppose that the rates of nonsynonymous and synonymous mutations differ. Mutations occur because of errors in DNA replication, and replication mechanisms do not sense how the alteration of a nucleotide affects protein coded for. The lower rate of nonsynonymous substitution

TABLE 8.2 **Synonymous and nonsynonymous substitution rates estimated by comparing genes in humans and mice**

Gene	Codons	Synonymous rate	Nonsynonymous rate
Histone *H3*	101	6.38	0.0
Histone *H4*	135	6.13	0.027
Growth hormone	189	4.37	0.95
Prolactin	197	5.59	1.29
α-hemoglobin	141	3.94	0.56
β-hemoglobin	144	2.96	0.87
γ-interferon	136	8.59	2.80
HPRT	217	2.13	0.13
Fibrogin-*γ*	411	5.82	0.55
Albumin	590	6.72	0.92

Source: Li et al. (1985). All rates are in units of 10^{-9} per site per year.

indicates that many of the nonsynonymous mutations fail to become fixed because they are deleterious. We can estimate the fraction that is strongly deleterious, α, by using the fact that the fixation probability of these mutations is 0. If we assume that the rest are neutral, the rate of fixation of nonsynonymous mutations is

$$r_N = (1 - \alpha)\mu \tag{8.4}$$

We find α for each gene by comparing the nonsynonymous rate with the average synonymous rate. For example, the nonsynonymous rate for *β-globin* is 0.8×10^{-9} per site per year. The average synonymous rate is 2.2×10^{-9} per site per year, which implies that $1 - \alpha = 0.8/2.2 = 0.36$ and hence $\alpha = 0.64$. That is, 64% of the nonsynonymous mutations in *β-globin* are strongly deleterious and 36% are neutral. This result is only approximate and does not allow for the possibility that some nonsynonymous mutations were advantageous or slightly deleterious, but it gives us a convenient summary of how selection constrains the evolution of the *β-globin* gene. We can say that at least 64% of the nonsynonymous mutations were deleterious. If some of them were only slightly deleterious, then even a larger fraction overall would have to have been deleterious. If some mutations were advantageous, then a still larger fraction would have to have been deleterious. Exercises 8.6 and 8.7 illustrate these points.

Most nonsynonymous substitution rates are less than synonymous rates, indicating that selection acts primarily to prevent the fixation of nonsynonymous mutations. Some nonsynonymous substitutions result from positive selection, but not enough to make the nonsynonymous rate higher than the synonymous

Figure 8.9 A dusky langur (*Trachypithecus obscurus*) feeding on acacia leaves.

rate. There are a few of exceptions, of course. An example is provided by the colobine monkeys, a group that includes colobus monkeys and langurs, such as the dusky langur (shown in **Figure 8.9**). Unlike other Old World monkeys, which primarily eat fruit and seeds, colobine monkeys eat large quantities of leaves. To allow them to survive on this unusual diet, colobines evolved a digestive system similar to that of cows and other ungulates. They have a second stomach (a foregut) in which digestive enzymes and symbiotic bacteria break down plant materials sufficiently to allow the extraction of nutrients that would otherwise be unavailable to them. One of the most abundant enzymes in the foregut of both colobines and ungulates is lysozyme, which in other mammals is found in saliva and tears and which protects against bacteria by rupturing their cell walls (lysing them, hence the name). On the lineage leading from other mammals to the ancestors of colobine monkeys, there are roughly nine times as many nonsynonymous substitutions as synonymous ones, which indicates that lysozyme on this linage experienced strong positive selection, presumably resulting from the new function it served in the ancestor of that group.

Genetic Hitchhiking

Hitchhikers go along for a ride—where the driver goes, the hitchhiker goes also. The same thing happens to neutral alleles at loci that are closely linked to a locus affected by selection. What happens to the neutral locus depends on what kind of selection is acting. If an advantageous allele is driven to fixation by positive selection, there is a *selective sweep*. Before an advantageous allele is fixed but after it increases substantially in frequency, there is a *partial sweep*. And if there is balancing selection, there is *associative overdominance* at linked neutral loci. Each of these processes creates characteristic patterns of variability at linked neutral loci, and detecting these patterns allows us to infer what kind of selection has been acting. We will describe each of these processes separately, in each one emphasizing the qualitative features of the patterns generated rather than the formal mathematical theory.

Selective Sweeps

Consider first the case of no recombination. All alleles on the chromosome on which the advantageous mutation occurs will be fixed when the ad-

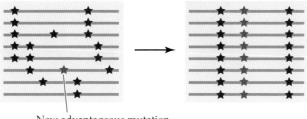

New advantageous mutation

Figure 8.10 The effect of a selective sweep in the absence of recombination. Each line in the figure represents a DNA sequence segregating in the population. Each star represents the derived allele of a SNP. A new advantageous mutation (red star) occurs with an initial frequency of 1/(2*N*) (left). After fixation of the advantageous alleles, all sequences are identical to each other (right).

vantageous allele is fixed, and all alleles not linked to the advantageous mutation will be lost. In the absence of recombination, the entire chromosome on which the advantageous mutation first occurred goes to fixation. This effect is illustrated in **Figure 8.10**. Notice that all genetic variation is eliminated from the region. There are no polymorphisms left because all alleles are either fixed or lost in the population. This effect is called a **selective sweep**. After the sweep is completed, new mutations will create new polymorphisms, and after a long time restore the previous level of polymorphism.

Now consider the case where recombination may occur. Chromosomal segments that otherwise would have been lost from the population may recombine with segments carrying the advantageous mutation. If this happens, not all polymorphisms on those segments will be lost from the population. They escape the sweep by recombining onto the new genetic background. This effect is illustrated in **Figure 8.11**.

The probability that recombination occurs between two sites depends on the recombination rate between them. Sites located very close to each

Escape by recombination

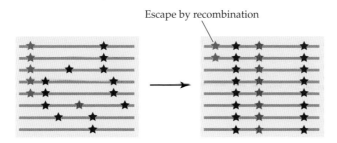

Figure 8.11 The effect of a selective sweep in the presence of recombination. Symbols are as in Figure 8.10, but the blue stars represent sites that have recombined onto chromosomes carrying the advantageous allele during the selective sweep.

Figure 8.12 Comparison of borzoi and boxer. The shorter muzzle of the boxer is the result of fixation of at least two mutations causing brachycephaly.

other in the sequence are less likely to recombine than are distant sites. As a consequence, the effect of the selective sweep will be more pronounced close to the selected site than farther away. A selective sweep will leave a characteristic pattern in the genome in which variability is low near the selected sites and is larger at more distant sites. Allele frequencies will also change in linked sites, but what happens depends on the distance to the selective sweep.

Two examples of selective sweeps have been found in the boxer, a breed of dog that has been strongly selected for having a shortened muzzle, a condition called brachycephaly, illustrated in **Figure 8.12**. On chromosomes 1 and 26, regions of very low heterozygosity surround two genes that have been shown to contribute to brachycephaly.

The formal theory of hitchhiking requires some algebra, which is presented in **Box 8.4** for a model of haploid organisms. From that theory, we can conclude the a selective sweep will have a strong effect on linked neutral

---

BOX 8.4 Hitchhiking in a Haploid Population

The theory of hitchhiking for a haploid population was developed by John Maynard Smith and John Haigh in 1972. For two loci with two alleles each, A/a and B/b, there are four haplotypes, AB, Ab, aB, and ab. A is the advantageous allele with selection coefficient s and the other locus is neutral. Let f_B be the frequency of B, let Q be the fraction of B-bearing chromosomes that carry A, and R be the fraction of b-bearing chromosomes that carry A. This notation is different from what we have used in other contexts but it simplifies the analysis of hitchhiking.

The haplotype frequencies and relative fitnesses are then:

Haplotype	AB	Ab	aB	ab
Frequency	f_{AB}	f_{Ab}	f_{aB}	f_{ab}
Fitness	1	1	$1-s$	$1-s$

BOX 8.4 *(continued)*

Haploid individuals mate at random, and recombination between the two loci occurs with probability c. The genotypes produced by each of the mating pairs and the frequency of that pair are: We find the haplotype frequencies in the next generation by adding over the sets of parents weighted by their combined fitnesses and dividing by the overall average fitness:

Parents		Frequency	Combined fitness	Offspring genotype			
				AB	**Ab**	**aB**	**ab**
AB	AB	f^2_{AB}	1	1	0	0	0
AB	Ab	$2f_{AB}f_{Ab}$	1	½	½	0	0
AB	aB	$2f_{AB}f_{aB}$	1–s	½	0	½	0
AB	ab	$2f_{AB}f_{ab}$	1–s	(1–c)/2	c/2	c/2	(1–c)/2
Ab	Ab	f^2_{bB}	1	0	1	0	0
Ab	aB	$2f_{Ab}f_{aB}$	1–s	c/2	(1–c)/2	(1–c)/2	c/2
Ab	ab	$2f_{Ab}f_{ab}$	1–s	0	½	0	½
aB	aB	f^2_{aB}	$(1-s)2$	0	0	1	0
aB	ab	$2f_{aB}f_{ab}$	$2(1-s)2$	0	0	½	½
ab	ab	f^2_{ab}	$(1-s)2$	0	0	0	1

$$f'_{AB} = \frac{f^2_{AB} + f_{AB}f_{Ab} + (1-s)f_{AB}f_{aB} + (1-s)f_{AB}f_{ab}(1-c) + (1-s)f_{Ab}f_{aB}c}{\bar{w}}$$

$$= \frac{f_{AB} - sf_{AB}f_a - c(1-s)D}{\bar{w}}$$

where $D = f_{AB}f_{ab} - f_{Ab}f_{aB}$ is the coefficient of linkage disequilibrium (Chapter 7) and $\bar{w}$ is the average fitness in the pairs,

$$\bar{w} = \left(1 - sf_a\right)^2$$

In a similar way,

$$f'_{Ab} = \frac{f_{Ab} - sf_{Ab}f_a + c(1-s)D}{\bar{w}}$$

and

$$f'_{aB} = \frac{(1-s)\left(f_{aB} - sf_{aB}f_a + cD\right)}{\bar{w}}$$

There is no simple solution to these equations, but they can be iterated for any initial condition. The case of interest is one in which one copy of A appears by mutation on a B-bearing chromosome when B has an initial frequency of $f_B(0)$, so initially $f_{AB} = 1/(2N)$, $f_{Ab} = 0$, $f_{aB} = f_B(0) - 1/(2N)$ and $f_{ab} = 1 - f_B(0)$. These are the equations used to generate the results shown in Figure 8.9.

Figure 8.13 Example of hitchhiking effect in a haploid model (see Box 8.4) with $s = 0.1$, $f_A(0) = 0.005$ and $f_B(0) = 0.2$. The solid lines are f_B and the dashed line is f_A.

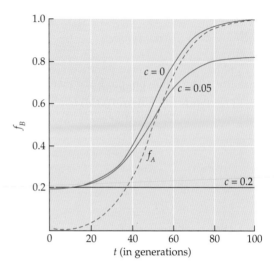

loci if the recombination distance, c, is less than the selection coefficient, s, in favor of the advantageous allele. **Figure 8.13** shows some numerical results for this model described in Box 8.4. Initially the neutral allele, B, is in frequency 0.2, and an initially rare advantageous allele with selection coefficient, $s = 0.1$ is driven to fixation. If $c = 0$, B is driven to fixation. If $c < s$, f_B is increased from 0.2 to roughly 0.8 and if $c > s$, f_B hardly changes.

Partial Sweeps

After an advantageous allele reaches an intermediate frequency but before it is fixed, there has been a **partial sweep**, in which some chromosomes carry the advantageous allele and have little or no variability among them, while other chromosomes do not have the advantageous allele and retain normal levels of polymorphism. A partial sweep also results in a characteristic pattern of polymorphism, illustrated in **Figure 8.14**. Polymorphism at sites closely linked to the advantageous allele is reduced, and a strong linkage disequilibrium is created between alleles on the same chromosome as the advantageous allele.

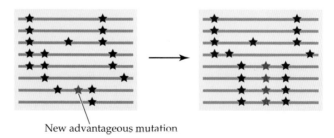

New advantageous mutation

Figure 8.14 An illustration of a partial selective sweep. The advantageous allele has increased in frequency, but not gone to fixation.

The pattern expected when there has been a partial sweep is similar to what we saw in the data presented in Figure 6.1, which showed the pattern of variation at sites linked to the *G6PD* gene in a West African population of humans. The major feature of that data set is the strong LD between the SNP creating the A^- allele and SNPs separated by some distance. In fact, the LD extends more than 700 kb from *G6PD*. The non-A^- chromosomes do not have LD on nearly that scale. The data show exactly what is expected if the A^- allele had recently increased in frequency because of directional selection. The size of the region of substantial LD allows us to conclude that the selection intensity in favor of A^- is at least 0.01 because selection of that intensity is required to produce substantial LD over such a large region of the chromosome. In fact, these data are from African individuals who live in a region where malaria is endemic. The A^- allele of *G6PD* is another example of an allele that has a selective advantage because it confers some resistance to the deadly effects of malaria. **Box 8.5** presents a simple method for estimating the time since a low frequency allele arose by mutation and a way to estimate the selection coefficient in favor of a rare mutation. We cannot determine whether the A^- allele will continue to increase in frequency because of the partial protection is provides from malaria or whether it has reached an equilibrium frequency under balancing selection of the type experienced by the *S* allele of the *β-globin* gene.

Associative Overdominance

Heterozygote advantage at a locus also affects closely linked neutral loci. We can understand why by considering a special case in which heterozygote advantage is so strong that it maintains two alleles at their equilibrium frequencies, $\hat{f}_A$ and $\hat{f}_a$. Those frequencies depend on the selection coefficients (Equation 7.9), but their numerical values will not be important.

Consider a neutral locus at recombination distance c from the selected locus. From the point of view of the neutral locus, there are actually two subpopulations, one being the *A*-bearing chromosomes and the other the *a*-bearing chromosomes. The size of the first subpopulation is $N\hat{f}_A$ and the size of the second subpopulation is $N\hat{f}_a$. Now consider what happens at the neutral locus when there is recombination. If recombination is in an individual who is homozygous for *A*, then the neutral site (denoted by *B* in **Figure 8.15**) will be linked to another copy of *A*. Therefore it will remain in

(A)   (B)

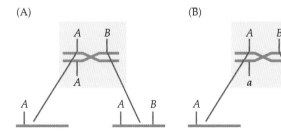

Figure 8.15 Illustration of the effect of recombination on a neutral site (*B*) linked to a site (*A/a*) subject to overdominant selection.

BOX 8.5 Estimating the Age of a Mutation

In Chapter 6, we defined D, the coefficient of linkage disequilibrium, as the difference between the observed haplotype frequency and the frequency expected if the two loci are independent: $D = f_{AB} - f_A f_B$. D characterizes the overall extent of nonrandom association between pairs of loci. When considering genetic hitchhiking, another measure is often useful. If A is the advantageous mutation that arose in a population previously fixed for a, then the extent of LD with a linked locus (B/b) can be characterized by δ, defined as the fraction of A-bearing chromosomes that also carry B:

$$\delta = \frac{f_{AB}}{f_A}$$

The reason for introducing δ is that if A first appears on a chromosome that already has a B, $\delta = 1$. While A is in low frequency, it will be in heterozygous individuals, and the effect of recombination on δ is easy to determine. Recombination affects δ only in AB/ab and Ab/aB double heterozygotes. When A is rare, the frequency of AB/ab individuals is approximately δf_b. Recombination in AB/ab individuals will reduce δ by c, the recombination rate between the two loci, because it will create Ab chromosomes. In a similar way, the frequency of Ab/aB individuals is approximately $(1-\delta)f_B$, and recombination will create AB chromosomes at a rate c. Therefore, δ in the next generation is approximately

$$\delta(t+1) = \delta(t) - cf_B\delta(t) + cf_b[1-\delta(t)]$$

Note that the change in δ depends only on the frequency of B and the recombination rate, not on the frequency of A. Furthermore, while A is rare, hitchhiking has not yet changed the frequency of B by much. Therefore, this equation has a simple solution:

$$\delta(t) = f_B + (1-f_B)(1-c)^t$$

Initially, $\delta = 1$. It decreases exponentially at a rate c to f_B.

If we know c and δ, we can solve for t, the time at which A arose by mutation:

$$t = \frac{\ln[(\delta-f_B)/(1-f_B)]}{\ln(1-c)} \tag{B8.1}$$

To illustrate, we use the data for the A^- allele of *G6PD* presented in Figure 6.1. At site 41 in the *IDH3G* locus, 13 of the 20 A^- chromosomes carry the derived allele. Therefore $\delta = 13/20 = 0.65$. Only one of the 31 A^+ and B chromosomes carries that allele, so the background frequency is $f_B = 1/31 = 0.026$. The loci are about 700 kb apart, which implies that $c = 0.007$ if 1 mb = 1cM. Therefore, we estimate t to be roughly 63 generations. If we assume the human generation time (i.e., the average age of a mother of a newborn) is 25 years, then we conclude that the A^- allele arose approximately 1600 years ago. A^- has increased to a frequency of 0.11 in western African populations in this short time, implying that it was favored by natural selection, probably because it confers some protection against malaria.

the same subpopulation. If, however, recombination is in an *Aa* heterozygote, *B* will have changed subpopulations because it will now be on an *a*-bearing chromosome. From the point of view of the neutral site, recombination plays the role of migration in a model with two subpopulations. The only difference from the model in Chapter 4 is that the subpopulation sizes are not equal and the "migration" rates between the two subpopulations are also not equal.

The effective migration rates in this model depend on the allele frequencies and the recombination rate, *c*. The probability that a neutral site linked to *A* will be linked to *a* after recombination is

$$m_{A \to a} = c\hat{f}_a \tag{8.5}$$

because $\hat{f}_a$ is the probability that an *A*-bearing chromosome will be in a heterozygous individual and *c* is the probability that recombination between the two loci will occur. Similarly,

$$m_{a \to A} = c\hat{f}_A \tag{8.6}$$

Without doing any calculations, we know that linkage to the selected site will increase the heterozygosity at the neutral site, because as we saw in Chapter 4 population subdivision makes the pairwise coalescence time at the neutral site longer than in a randomly mating population. How much longer and hence how much more heterozygosity will result depends on the recombination rate.

The calculation of the average coalescence times for the general case is complicated and requires the theory of Markov chains, which is beyond the scope of this book. For the special case in which $\hat{f}_A = \hat{f}_a = \frac{1}{2}$, this problem becomes equivalent to the model of two subpopulations, which you have already seen in Chapter 4. In terms of the parameters in Chapter 4, the population size, *N*, is half the population size in this model, and $m = c/2$. On page 68 of Chapter 4, you found that the average coalescence time of two copies drawn from the same population is 4*N*, so in this model the average coalescence time of two copies on the same type of chromosome (*A*- or *a*-bearing) is 2*N*. The average coalescence time of two copies that are initially in difference subpopulations is $2N + 1/m$, so in this model, the average coalescence time of two copies initially on different types of chromosomes is $2N + 2/c$. The expected heterozygosity per site at the neutral locus is the average of these two quantities, H_T on page 69 of Chapter 4:

$$H_T = \left(\frac{1}{4Nc} + 1\right)\frac{\theta}{k} \tag{8.7}$$

where θ is the mutation rate multiplied by 2*N* and *k* is the number of sites. You can see that H_T increases with $1/c$. That is true when $\hat{f}_A$ is not ½, as well.

If there is balancing selection affecting a locus, what we would expect to see is a local increase in heterozygosity near that locus. By contrast, if there is directional selection at a locus, we would see a local decrease. Examples

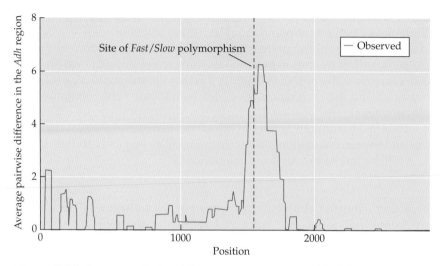

Figure 8.16 Average pairwise difference in sequence in 1-kb sliding windows between *Fast* and *Slow* alleles of the *Adh* gene in *D. melanogaster*. The data are consistent with heterozygote advantage of the site that distinguishes *Fast* from *Slow*. The dashed line indicates the site of the *Fast*/*Slow* polymorphism (site 1552). Position is the number of nucleotides from the end of the *Adh* region sequenced. (After Hudson and Kaplan, 1988.)

of both kind of patterns are known, but the evidence so far suggests that areas of reduced heterozygosity caused by selective sweeps are much more common than areas of increased heterozygosity caused by balancing selection. One notable example is the *Adh* locus in *Drosoplila melanogaster*. Two alleles, called *Slow* (*S*) and *Fast* (*F*) appear to be maintained by balancing selection. As expected, heterozygosity near the *Adh* is higher than in other parts of the *D. melanogaster* genome (**Figure 8.16**).

In summary, selection affecting a locus also affects nearby neutral loci, either decreasing or increasing heterozygosity. This hitchhiking effect lets us identify regions whether selection has occurred, as will be described in Chapter 9.

References

Hästbacka J., de la Chapelle A., Kaitila I. et al., 1992. Linkage disequilibrium mapping in isolated founder populations: diastrophic dysplasia in Finland. *Nature Genetics* 2: 204–211. http://www.nature.com/ng/journal/v2/n3/abs/ng1192-204.html

Hudson R. R. and Kaplan N. L., 1988. The coalescent process in models with selection and recombination. *Genetics* 120: 831–840. http://www.genetics.org/cgi/reprint/120/3/831

Li W. H., Wu C. I., Luo C. C., 1985. A new method for estimating synonymous and nonsynonymous rates of nucleotide substitution considering the relative likelihood of nucleotide and codon changes. *Molecular
Biology and Evolution* 2: 150–174.

Maynard Smith J. and Haigh J., 1974. The hitch-hiking effect of a favour-
able gene. *Genetical Research* 23: 23–35. http://journals.cambridge.org/
action/displayAbstract?fromPage=online&aid=1754360&fulltextType
=RA&fileId=S0016672300014634

Risch N., de Leon D., Ozelius L., et al., 1995. Genetic analysis of idiopathic torsion dystonia in Ashkenazi Jews and their recent descent from
a small founder population. *Nature Genetics* 9: 152–159. http://www.
nature.com/ng/journal/v9/n2/abs/ng0295-152.html

Yang, Z. H., 1998. Likelihood ratio tests for detecting positive selection
and application to primate lysozyme evolution. *Molecular Biology
and Evolution* 15, 568–573. http://mbe.oxfordjournals.org/con-
tent/15/5/568

EXERCISES

8.1 Suppose that the mutation rate for single nucleotide change is 2.2×10^{-9} per site per year.

   a. What is the rate of substitution of deleterious mutations per million years if the selection coefficient against them is 0.001 in a
   population containing 10,000, 1000, or 100 individuals?

   b. What fraction of the neutral rate are the substitution rates you
   computed for part a?

8.2 Find the same results as in Exercise 8.1 for an advantageous mutation with a selection coefficient 0.001.

8.3 The fixation probability for a recessive advantageous mutant
$(v_{AA} = 1, v_{Aa} = v_{aa} = 1 - s)$ in a population containing N individuals is
$\sqrt{2s/(N\pi)}$. Notice that this probability depends on N no matter
how large s is. The fixation probability of a strongly advantageous
allele with an additive effect of 0.1% on viability is approximately $2s$
= 0.002. How large would s have to be for a recessive advantageous
allele to have the same fixation probability in a population of 10,000
individuals?

8.4 Suppose you compare 1000 codons of aligned sequence in humans
and chimpanzees. You do not have a computer available, so you examine the second codon positions by hand and find four that differ.
What is your estimate of the nonsynonymous substitution rate?

8.5 For insulin, the rate of nonsynonymous substitutions is 0.13×10^{-9} per site per year, and for histones, it is about 10^{-13} per site per year. What is the minimum fraction of the nonsynonymous deleterious mutations that are deleterious if the neutral rate is 2.2×10^{-9} per site per year?

8.6 The rate of synonymous substitution for *β-globin* is 0.8×10^{-9}. In the text, we showed that if a fraction α of the nonsynonymous mutations is strongly deleterious and the rest are neutral, $\alpha = 0.64$. Suppose instead that a fraction α of the nonsynonymous mutations are deleterious but that only ⅔ of those are strongly deleterious. The remaining ⅓ are slightly deleterious with a selection coefficient of 0.001. Use the results from Exercise 8.1 to find α. Assume that the population size is 100 and that the mutation rate is 2.2×10^{-9} per site per year.

8.7 Assume that for *β-globin*, 0.08% of the nonsynonymous mutations are strongly advantageous with a selection coefficient of 0.01 in a population of size 10,000, and that a fraction α are strongly deleterious. If the mutation rate is 2.2×10^{-9} per site per year, what is α?

8.8 Idiopathic torsion dystonia (ITD) is a movement disorder caused by dominant alleles at the *ITY* locus on chromosome 9 in humans. In the Ashkenazi Jewish population, most cases of ITD are caused by a single mutation. On 54 chromosomes with this mutation, 47 have a particular allele (12) at a microsatellite locus (*ASS*) that is 2.3 cM away from *ITY*. The background frequency of allele 12 at *ASS* is 0.086. How many generations in the past did the causative mutation arise? (Use Equation B8.1 in Box 8.5.)

8.9 Disastrophic dysplasia (DTD) is a disorder that causes short stature and unusual growth of the joints in humans. Many cases are caused by dominant alleles at a locus on chromosome 5. In a study of 146 chromosomes from the population of Finland, 139 were found to have an allele at the *CSF1R* locus that is in 3% frequency on non-DTD chromosomes. Estimate the recombination rate between *CSF1R* and the locus that causes DTD under the assumption that the Finnish population was founded 100 generations ago and that one of the founders carried the mutation causing DTD. (Use Equation B8.1 in Box 8.4.) This method for estimating the recombination distance is called linkage disequilibrium mapping.

8.10 Suppose that ten alleles at a gametic self-incompatability locus are present in a plant population in equal frequencies. The model is equivalent to an island model of population structure with ten islands.

    a. What is the effective migration rate in this model? (Remember, there are no homozygotes at the S locus.)

    b. Use the theory in Chapter 4 for an island population with d demes to predict the expected heterozygosity at a neutral locus at a recombination distance c from the S locus. You have to calculate H_T by averaging the probability that two copies of the neutral locus are on the same S allele background and on different backgrounds.

    c. Would you expect the heterozygosity of the neutral locus to increase or decrease if the number of S alleles were larger than 10?

9 The Neutral Theory and Tests of Neutrality

ONE OF THE KEY ASSUMPTIONS of Darwin's theory of natural selection was that there was heritable variation in all phenotypic characters. In the twenty years after the rediscovery of Mendelian inheritance in 1900, biologists demonstrated that heritable phenotypic variation could be attributed to Mendelian genes. After that time, the theory of population genetics was developed in part to show that natural selection affecting Mendelian genes could account for short-term and long-term evolution, thus creating part of the foundation of the neo-Darwinian synthesis.

Before the emergence of DNA sequence data, it was very unclear how much genetic variation existed in natural population and how much of this variation was affected by natural selection. However, soon after the first molecular data appeared, quantification of the relative contributions of natural selection and genetic drift to genetic differences within and between species became one of the dominant goals for researchers of molecular evolution. In 1962, using amino acid sequences, E. Zuckerkandl and L. Pauling discovered the molecular clock discussed in Chapter 2. The population geneticist M. Kimura then realized that the existence of a molecular clock could be explained by a combination of mutation and genetic drift, and that no selection was needed to understand the accumulation of genetic differences between species. This led him, in 1968, to propose the **neutral theory of molecular evolution**, which posits that molecular variation within and between species can be explained by mutation and genetic drift alone, without the action of selection. The theory was later modified to include some forms of weak selection.

Researchers in molecular evolution divide selection on new mutations into two categories. **Positive selection** is selection acting in favor of new mutations, in other words, selection in which the new mutation is associated with a positive selection coefficient. **Negative selection** refers to the opposite case, in which selection acts against new mutations. The **neutral theory** allows for the presence of strongly deleterious mutations, that is, mutations affected by strong negative selection, because new mutations affected by this type of selection will not contribute to variation within and between species. If selection is sufficiently strong, such mutations will never segregate in the population. However, the main tenet of the neutral theory is that strongly positively selected mutations play only a very minor role in molecular evolution.

The division of selection into positive and negative selection applies to new mutations. But because environmental conditions are subject to change, selection affecting alleles already segregating in the population may change and possibly favor the ancestral allele instead of a new mutation. The previous definition of positive and negative selection clearly does not apply in such cases. Also, selection may change through time so that selection acts against an allele that selection previously acted for, and vice versa. Again, the semantics of positive and negative selection do not apply in these cases. However, the neutral theory assumes that almost all selection acts in favor of alleles that are very common in the population, and against rare mutations. So scenarios with changing selection coefficients are not encompassed by the neutral theory.

The formulation of the neutral theory led to several decades of arguments among scientists regarding its validity. The arguments have faded today, as most research instead focuses on detecting, describing, and understanding specific instances of selection. However, the neutral theory is still central to our understanding of molecular evolution and plays an important role as a null hypothesis used in studies aimed at detecting natural selection.

The typical approach for detecting natural selection is to use statistical tests that examine whether a neutral model fits the observed data. If the neutral model does not fit, natural selection is invoked. We have already discussed one test of neutrality: the comparison of the rate of nonsynonymous and synonymous mutation (Chapter 8). We saw that most genes had a lower rate of nonsynonymous than synonymous mutations, leading to the conclusion that negative selection is affecting these genes. Such comparisons are often formalized in statistical tests. In these tests, the rate of nonsynonymous substitutions per nonsynonymous sites (d_N) is compared to the rate of synonymous substitutions per synonymous sites (d_S). As discussed in Box 8.3, estimation of d_N and d_S can be complicated, but we have powerful statistical methods for estimating d_N and d_S that can take the intricacies of the genetic code into account. The ratio of d_N/d_S is then used to make inferences about selection. Negative selection is inferred when $d_N/d_S < 1$, and positive selection when $d_N/d_S > 1$; $d_N/d_S = 1$ is compatible with neutrality. Researchers are particularly interested in identifying positive selection because it provides evidence of adaptation at the molecular level.

Figure 9.1 A cartoon structure of the influenza hemagglutinin molecule. This molecule is located on the surface of the viral capsid (the protein shell of the virus) and is the primary target of the immune system. One of the reasons humans may be affected by many influenza infections during their lifetimes is that this molecule evolves very fast. Immunity to one variety of influenza may not provide immunity to other newly evolved versions of the influenza virus carrying different hemagglutinin molecules. Phylogenetic comparisons of multiple hemagglutinin DNA sequences reveal that some of the sites evolve with values of d_N/d_S vastly larger than 1. The amino acid residues corresponding to sites identified in one analysis to have evolved with $d_N/d_S > 1$ are shown as red bubbles in the figure. These residues are known to be targets of the human immune system. It is believed that the accelerated rate of nonsynonymous substitution in this gene is caused by the selection exerted on the virus by the host immune/defense system.

While d_N/d_S originally was applied to pairs of sequences, it is now routinely applied simultaneously to multiple sequences using a phylogenetic tree, and estimation of d_N/d_S can be done for a subset of sites or for specific branches of the phylogenetic tree. For most genes, d_N/d_S is less than 1. Many nonsynonymous mutations are deleterious, leading to reduced values of d_N/d_S. The average d_N/d_S, therefore, is rarely if ever statistically significantly larger than 1 when averaging over many sites. However, subsets of sites may still be affected by positive selection and evolve with $d_N/d_S > 1$, even when the average d_N/d_S for all sites in a gene is less than one. Fortunately, there are a number of statistical methods that allow researchers to identify sites affected by positive selection located among a larger proportion of sites dominated by negative selection. An example is given in **Figure 9.1**.

If enough mutations have been affected by positive selection for the sequences to incur a statistically significant elevation of the d_N/d_S ratio, this ratio allows us to detect positive selection in comparisons of data from different species, or from different strains of viruses or bacteria. This was illustrated in the evolution of lysozyme in the lineage leading to colobine monkeys (Chapter 8). However, if selection has acted on a single mutation, or has been affecting a noncoding region, methods based on d_N/d_S have no power to detect positive selection. It is often of great interest to determine if selection has acted in the recent history of a population, perhaps affecting just one or a few sites in a sequence. To detect this type of selection, it is not sufficient just to compare DNA sequences from different species. Analyses must instead be based on multiple sequences from different individuals from the same population, i.e., population genetic data. In the next section, we will describe a number of different tests of neutrality that routinely are used to detect natural selection using population genetic data.

The HKA Test

One of the first tests proposed to detect natural selection from DNA sequence data is the **HKA test**, named after population geneticists R. R. Hudson, M. Kreitman, and M. Agaudé. This test is based on comparing variability within and between species.

In Chapter 3 we saw that, assuming an infinite sites model, the expected number of segregating sites within a species is $E[S] = \theta \sum_{i=1}^{n-1} 1/i$, where n is the sample size (number of chromosomes) and $\theta = 4N\mu$, where N is the population size and μ is the mutation rate per generation. The rate of substitution between species equals μ, so ignoring ancestral variation, the number of fixed differences between species is $2T\mu$, if T is the divergence time measured in number of generations. If we take variation in the ancestral population into account, the number of differences between two sequences sampled from different species is $2T\mu + 4N_A\mu$, where N_A is the population size of the ancestral population. We simply add the number of mutations in the ancestral population to the ones expected after the divergence of the two species. Notice that the expected number of mutations, both within and between species, is proportional to the mutation rate. So the ratio of the expected number of mutations within and between species depends not on the mutation rate, but on the effective population sizes, the sample size, and the divergence time. In data of multiple loci with the same sample size, from the same species, the ratio of the number of fixed differences between species to the number of segregating sites within species is, therefore, expected to be the same for all loci. Consider, for example, two loci with data as in **Table 9.1**, then under neutrality, $E[S_1]/E[F_1] = E[S_2]/E[F_2]$.

This result is also true under demographic models other than the standard coalescence model. If the ratios are very different from each other, this may indicate that selection is acting on one of the loci. For example, if S_2/F_2 is much smaller than S_1/F_1, it could be because a selective sweep has recently affected locus 2 and reduced the number of segregating sites in this locus. But there could also be other explanations. For example, balancing selection might be affecting locus 1. When analyzing genome-wide data, searching for loci with reduced levels of variability is a common method for finding genes that recently have been affected by selective sweeps.

Because of linkage disequilibrium between SNPs within each locus, it is often difficult to find critical values for the HKA test. A simple chi-square

TABLE 9.1 An example of an HKA table

	Locus 1	Locus 2
Segregating sites within species	S_1	S_2
Fixed differences between species	F_1	F_2

TABLE 9.2   HKA tests for two introns from the *Dmd* locus

Geographic region	Locus	S	F	p value
Africa	Intron 7	6	39	*NS*
	Intron 44	15	27	
Europe	Intron 7	1	39	<0.05
	Intron 44	10	27	
Asia	Intron 7	0	39	<0.01
	Intron 44	10	27	
Americas	Intron 7	3	39	*NS*
	Intron 44	9	27	

Note: S is the number of segregating sites (SNPs) and F is the number of fixed differences between humans and chimpanzees. *NS* means "not significant."
Source: Nachman and Crowell (2000).

test is not appropriate; instead, population geneticists rely on coalescence simulations (see Chapter 5).

In the HKA test shown in **Table 9.2**, Nachman and Crowell were investigating whether selection has been affecting genetic variation in two introns of the *Duchenne muscular dystrophy* (*Dmd*) gene in humans. To obtain fixed differences, they compared the human sequence to the chimpanzee sequence. They found a strong reduction in variability in intron 7 compared to intron 44 for several of the populations investigated. They tested for statistical significance using an HKA test, and concluded that it is likely that a selective sweep in Asia and Europe has reduced the amount of variability in the region around intron 7.

The MacDonald–Kreitman (MK) Test

Another popular test of neutrality is the **MacDonald–Kreitman (MK) test**. This test is similar to the HKA in that it compares variation within and between species. However, it does so by dividing mutations in coding regions into synonymous and nonsynonymous mutations, as in the d_N/d_S based tests.

If all segregating or fixed mutations are neutral, then the proportion of fixed differences that are nonsynonymous should be the same as the proportion of segregating mutations that are nonsynonymous. In terms of the notation in **Table 9.3**, we would expect $E[S_N]/E[S_S] = E[F_N]/E[F_S]$, or, similarly, $E[S_N]/E[F_N] = E[S_S]/E[F_S]$. If the observed value of S_N/S_S is much different from F_N/F_S, this provides evidence against neutrality and in favor of selection. More specifically if $F_N/F_S > S_N/S_S$, this provides evidence for positive selection; if $F_N/F_S < S_N/S_S$, it provides evidence of negative selection (or perhaps balancing selection), at least if the populations analyzed

TABLE 9.3 A MacDonald–Kreitman (MK) table

	Non-synonymous	Synonymous
Segregating sites within species	S_N	S_S
Fixed differences between species	F_N	F_S

have been of constant size through time. Under some assumptions, negative selection may result in $F_N/F_S > S_N/S_S$ if population sizes have increased through time. Also, $F_N/F_S < S_N/S_S$ may be a result of positive selection for declining populations.

To test if $F_N/F_S = S_N/S_S$, a chi-square test can be used similarly to the test for *LD* described in Box 6.4.

The first genome-wide scan for selected loci in the human genome was carried out by C. Bustamante and colleagues in 2005 using genome-wide sequencing data from humans. By performing an MK test for each locus, they constructed a genome-wide map of loci under positive and negative selection in humans. While some loci were found to be targets of positive selection, vastly more loci provided evidence of negative selection. The MK test does not detect selection acting on strongly deleterious mutations, because such mutations will not segregate in the population; they will immediately be eliminated. The negative selection detected by Bustamante and colleagues is, therefore, selection acting on segregating mutations that are slightly or moderately deleterious. Genes found to be under negative selection tend to be associated with basic cellular processes, and include genes coding for proteins such as cytoskeletal proteins. Genes associated with positive selection include many genes involved in the immune and defense system, such as T-cell receptors. Not surprisingly, just as viruses evolve to adapt to the host immune system, so does the host genome, in response to the selective pressures imposed by the constantly changing pathogenic environment.

The Site Frequency Spectrum (SFS)

We introduced the Site Frequency Spectrum (SFS) in Chapter 3 and discussed in Chapter 5 how it can be used for making inferences regarding the demographic history of the populations. However, as selection acts to change allele frequencies, the SFS also provides information regarding selection.

If new mutations are affected by negative selection, they will tend to segregate in the population at lower frequencies. As a consequence, the SFS for negatively selected alleles will be skewed toward low-frequency alleles compared to the distribution expected for neutral alleles (**Figure 9.2**).

Positive selection will have the opposite effect. Positively selected alleles will tend to segregate at higher frequencies, so the SFS will be skewed toward higher allele frequencies (Figure 9.2).

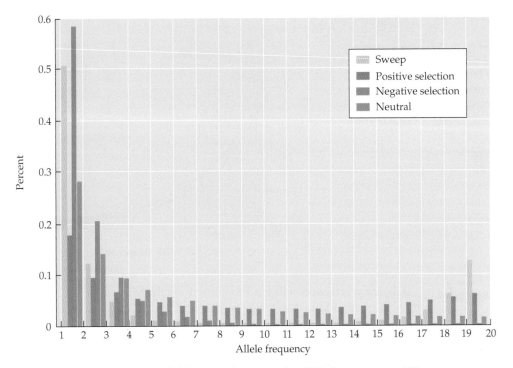

Figure 9.2 The expected (unfolded) SFS for a sample of 20 chromosomes (10 diploid individuals) under different models: the standard neutral model (neutral); a model in which negative selection is acting independently on each new mutation with a scaled selection coefficient of $2Ns = -5$ (negative selection); a model in which positive selection is acting independently on each new mutation with a scaled selection coefficient of $2Ns = 5$ (positive selection); and a model in which a positively selected mutation has just reached fixation (sweep). The SFS in the sweep model depicts the distribution of allele frequencies of the neutral mutations in a region around the site of advantageous mutation. (After Nielsen, 2005.)

However, as previously discussed, when selection acts to change the allele frequency of advantageous or deleterious mutations, the allele frequencies of linked mutations will also be affected. For example, a selective sweep will cause strong changes in the allele frequencies of mutations linked to the advantageous mutations. Figure 9.2 shows the expected frequency spectrum under one model of a selective sweep, for the neutral mutations in a genomic region surrounding the selected site. Notice that there is an excess of both low- and high-frequency alleles compared to that expected under neutrality. The alleles that were not linked to the advantageous mutation have decreased in frequency, and the alleles that were linked to the advantageous allele have increased in frequency. As a consequence, there are excesses of high-frequency and low-frequency alleles, but many fewer alleles of intermediate frequency.

186

Figure 9.3 The SFS in a sample of 200 humans from Denmark in a collection of approximately 15,000 protein-coding genes. The SFS has been calculated separately for nonsynonymous (NSyn) and synonymous (Syn) mutations. The SFS expected under a standard neutral model (Neutral exp.) is also shown.

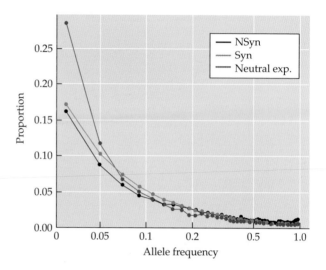

Balancing selection will tend to show the opposite pattern. Linked alleles will be maintained in the population at higher frequencies than that expected under the standard neutral model, and the SFS will contain more alleles of intermediate frequency.

In real data the effect of selection on allele frequencies is easily discernible. In human data, and in data from most other species, the SFS for nonsynonymous mutations is skewed toward more rare alleles than the SFS for synonymous mutations (**Figure 9.3**). This pattern is thought to be caused by negative selection acting on nonsynonymous mutations. Under the assumption of independence among sites, the amount of selection required to cause this pattern can be calculated using methods that are beyond the scope of this book. Depending on the particulars of the assumptions, estimates of the proportion of nonsynonymous segregating mutations affected by negative selection in the human genome varies from 40–80%. If linkage is taken into account, the proportion might be even higher.

A number of statistical tests use the SFS to detect evidence for selection. Most of these tests are designed to detect the effect of a selective sweep by examining whether the observed distribution of SNP allele frequencies in a region fits that expected under a standard neutral model. The most famous of these is Tajima's *D* test, discussed in the next section.

Tajima's *D* Test

The tests discussed so far all rely on simply counting mutations and dividing them into different categories. However, as discussed in the previous section, allele frequencies are very informative regarding selection. **Tajima's *D* test** is constructed to detect deviations in the distribution of allele frequencies from that expected under the standard neutral model.

Recall from Chapter 3 the two unbiased estimators of θ under the infinite sites model: Watterson's estimator, $\hat{\theta}_W$, based on the number of segregating sites; and Tajima's estimator, $\hat{\theta}_T$, based on the average number of pairwise differences:

$$\hat{\theta}_W = S \bigg/ \left(\sum_{i=1}^{n-1} 1/i \right) \text{ and } \hat{\theta}_T = \frac{\sum_{(i,j):i<j} d_{ij}}{n(n-1)/2} \tag{9.1}$$

where n is the number of chromosomes sampled, S is the number of segregating (variable) sites and d_{ij} is the number of nucleotide differences between sequence i and j in the sample. As discussed in Chapter 3, these two estimators are expected to provide the right estimate when the standard neutral model with infinite sites mutation is correct. However, if the two estimates are very different from each other, it may suggest that this model is not correct. One of the reasons this model may not be correct is that selection has been acting. We saw that a selective sweep will generate an SFS with many more common and rare derived alleles than expected under the standard neutral model. Mutations that are very rare or very common contribute less to the average number of pairwise differences than mutations that are of intermediate frequency. A selective sweep, therefore, has a stronger effect on $\hat{\theta}_T$ than on $\hat{\theta}_W$. In a region affected by a recent selective sweep, we would expect $\hat{\theta}_W$ to provide a larger estimate of θ than $\hat{\theta}_T$. Balancing selection will result in the opposite pattern. Each mutation will contribute more to $\hat{\theta}_T$ than expected under neutrality, and $\hat{\theta}_T$ will tend to be larger than $\hat{\theta}_W$. Tajima, therefore, suggested using

$$\text{Tajima's } D = \frac{\hat{\theta}_T - \hat{\theta}_W}{\sqrt{\hat{V}(\hat{\theta}_T - \hat{\theta}_W)}} \tag{9.2}$$

as a test statistic when testing the neutral infinite sites model. $V(\hat{\theta}_T - \hat{\theta}_W)$ is the variance (see Appendix A) in $(\hat{\theta}_T - \hat{\theta}_W)$ and is, therefore always positive. As a mathematical symbol, D has also been used in this book as a symbol for the linkage disequilibrium coefficient. To avoid confusion, we will therefore always use "Tajima's D" in the present context.

We would expect Tajima's D to be (approximately) equal to zero under the standard neutral model, to have a negative value for data sampled right after a selective sweep, and to have a positive value in the presence of balancing selection. Critical values for a test based on this test statistic are obtained using coalescence simulations (see Chapter 5); they cannot be calculated directly in any simple way. As a very rough approximation, in the absence of recombination, Tajima's D test will reject at the 5% significance level if Tajima's $D > 1.8$ or Tajima's $D < -1.8$. However, to obtain more precise p values, simulations must be used that take into account the local recombination rate.

Tajima's D test has been used extensively in humans and other organisms to detect selection. One example is the analyses of the *FOXP2* gene. Humans with mutations in the *FOXP2* gene have difficulties forming grammatically correct sentences, but they do not otherwise suffer from any physical or mental disabilities. Researchers therefore thought that this gene might be related to the evolution of human speech. To investigate this further, researchers compared the *FOXP2* gene from many different mammalian species, and found that the gene was highly conserved. At the amino acid level, almost all species were identical. However, surprisingly, humans had two unique nonsynonymous mutations in this gene. To test more directly if the gene might have experienced a recent selective sweep, they used Tajima's *D* test. They sequenced more than 14,000 bp of DNA around the gene from each of 20 individuals. They found 47 SNPs and estimates of $\hat{\theta}_W = 7.9 \times 10^{-4}$ and $\hat{\theta}_T = 3.0 \times 10^{-3}$. As expected for a region targeted by a selective sweep, $\hat{\theta}_W > \hat{\theta}_T$. The value for Tajima's *D* was -2.20, which was found to be significant at the 1% significance level using coalescence simulations. The researchers concluded that the gene region has indeed been the target of a selective sweep in recent human history.

Tests Based on Genetic Differentiation among Populations

So far we have discussed selection acting in a single population. But as discussed in Chapter 4, most natural populations are structured, consisting of multiple subpopulations that may differ in their allele frequency distributions. We have seen that under simple models of mutation, genetic drift and migration between populations, substantial allele frequency differences may exist if the number of migrations is less than, say, one per generation. Such differences in allele frequencies can be strengthened by selection for several reasons. First, selection may act differently in different populations, leading to long-lasting differences in allele frequencies between the populations at the loci affected by selection. Second, positive selection may lead to increased differences in allele frequencies in a transient period during which the frequency of the selected allele is higher in the population in which the mutation first arose. However, it is important to realize that selection could also lead to more similar allele frequencies than expected without selection, for example, if the same type of balancing selection is acting in all populations. Even directional positive selection may, under some assumptions, lead to more similar allele frequencies among populations. Nonetheless, identifying strong differences in allele frequencies has been one of the most useful tools for detecting adaptation in a local population.

An example is the *lactase* (*LCT*) locus in humans. The lactase enzyme encoded by this locus breaks the milk carbohydrate lactose into glucose and galactose. All healthy human infants produce this enzyme, which allows them to digest their mother's milk. However, the expression of the *LCT* is reduced in many adults, leading them to become lactose intolerant.

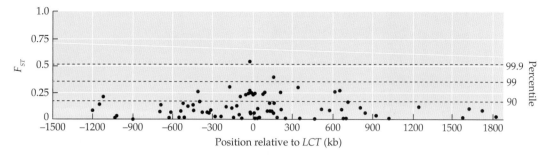

Figure 9.4 Estimates of F_{ST} in a region around the *lactase* (*LCT*) locus for a sample of Europeans, Africans, and Asians. The percentile is among all SNPs in the genome. (After Bersaglieri et al., 2004.)

Individuals who keep expressing this gene as adults, allowing them to digest milk through their entire life, are said to have "lactase persistence." The frequency of lactose intolerance varies greatly between geographic regions. In Denmark and Sweden, the frequency of lactose intolerance among adults is only 1 or 2%; however, in Thailand, it is above 97%. Researchers have determined that lactose persistence/intolerance in Europe is mostly determined by a single mutation in the upstream regulatory region of the *lactase* gene. Almost all Europeans with lactase persistence are either heterozygous or homozygous for this mutation, and almost all Europeans with lactose intolerance are homozygous for the alternative allele. It is thought that the allele leading to lactose persistence has been under positive selection in Europe, particularly northern Europe, since the emergence of dairy farming almost 10,000 years ago. Individuals with the ability to digest cow's milk would have had increased fitness in regions with dairy cattle farming, and especially so in Northern Europe, where food typically would have been scarce during the winter. One of the clear signals that the *lactase* locus has been under selection is its unusually high F_{ST} (**Figure 9.4**). F_{ST} is a measure of allele frequency differences between populations (see Chapter 4). F_{ST} between European and other continental populations is > 0.5 in a region around the *LCT* locus, despite the fact that F_{ST} is quite small for human populations in general.

Another example is the *EPAS1* locus in Tibetans. The two-dimensional SFS for Han Chinese and Tibetan individuals was shown in Figure 5.14 As discussed in the figure legend, the allele frequencies of Han Chinese and Tibetans are highly correlated, suggesting that these two populations are closely related. However, a few SNPs do not follow this pattern. In particular, the two red dots in the lower right of Figure 5.14 represent two SNPs that both have an extreme pattern of low frequency in Han Chinese and high frequency in Tibetans. Both of these SNPs are in the *EPAS1* gene. This gene shows a very strong and unusual pattern of high differences in allele frequencies between Han Chinese and Tibetans. Interestingly, the

Figure 9.5 Tibetans provide a remarkable example of humans adapting evolutionarily to local environmental conditions, in this case the hypoxic environment of the Tibetan plateau.

function of this gene is to regulate the response to hypoxia, that is, low oxygen levels. Most Tibetans live on the Tibetan plateau at an altitude of more than 4000 meters. At this high altitude, the oxygen pressure is 40% less than at sea level. It is thought that the allele frequency differences between Tibetans and Han Chinese in this gene are caused by selection acting on the Tibetan population in relation to altitude adaptation. In fact, there is a correlation between the *EPAS1* genotypes in the highly differentiated SNPs and hemoglobin levels in Tibetan individuals, suggesting a direct functional link between *EPAS1* and altitude adaptation among Tibetans (**Figure 9.5**).

Tests Using LD and Haplotype Structure

When discussing a selective sweep, we have so far assumed that the main interest is in detecting a complete sweep, in other words, a selective sweep in which the advantageous allele has reached a frequency of 100% in the population. However, it is perhaps even more interesting to be able to detect ongoing selection, i.e., selection still acting on a mutation that is segregating in the population. Some of the methods we have discussed can detect such selection. For example, the SFS will be affected by selection before the selected mutation reaches fixation, so tests based on the SFS, such as Tajima's D, can detect this type of selection. However, as discussed in Chapter

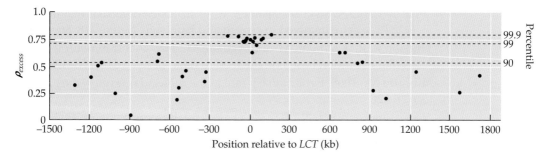

Figure 9.6 Estimates of ρ_{excess} in a region around the *lactase* (*LCT*) locus for a sample of Europeans, Africans, and Asians. ρ_{excess} is a measure of the increase in linkage disequilibrium. Notice the marked increase in the selected region around the *LCT* locus. The percentile is among all SNPs in the genome. (After Bersaglieri et al., 2004.)

8, a major effect of positive selection is to increase the level of LD in the population. The effect of a partial selective sweep on haplotype structure in a region around the site of the advantageous mutation is illustrated in Figure 8.14. As the advantageous mutations reach intermediate frequencies, a large proportion of the chromosomes segregating in the population will be identical in this region. However, the chromosomes not carrying the advantageous allele will have a haplotype structure similar to the one observed before the advantageous allele arose. This is a very specific pattern that cannot easily be generated by processes other than selection.

Several statistical tests have been designed to detect this type of pattern. They are all based on the same fundamental insight: in the presence of an incomplete sweep, many haplotypes in the region affected by the sweep should be nearly identical (have low haplotype homozygosity), while a fraction of the chromosomes when compared to each other will have normal levels of haplotype homozygosity. An example of the increase in linkage disequilibrium caused by an incomplete selective sweep is shown in **Figure 9.6**.

Identification of natural selection remains one of the most important applications of population genetic theory. Examples such as the selection on *EPAS1* in Tibetans, or on the influenza hemagglutinin molecule discussed in this chapter, illustrates the use of population genetics, not only for understanding evolution, but also for elucidating functional relationships.

References

*Bersaglieri T., Sabeti P. C., Patterson N., et al., 2004. Genetic signatures of strong recent positive selection at the lactase gene. *American Journal of Human Genetics* 74: 1111–1120.

*Bustamante C. D., Fledel-Alon A., Williamson S., et al., 2005. Natural selection on protein-coding genes in the human genome. *Nature* 437: 1153–1157.

*Enard W., Przeworski M., Fisher S. E., et al., 2002. Molecular evolution of *FOXP2*, a gene involved in speech and language. *Nature* 418: 869–872.

Hudson R. R., Kreitman M. and Aguadé M., 1987. A test of neutral molecular evolution based on nucleotide data. *Genetics* 116: 153–159.

*Kimura M., 1968. Evolutionary rate at the molecular level. *Nature* 217: 624–626.

*Li Y., Vinckenbosch N., Tian G., et al., 2010. Resequencing of 200 human exomes identifies an excess of low-frequency non-synonymous coding variants. *Nature Genetics* 42: 969–972.

*McDonald J. H. and Kreitman M., 1991. Adaptive protein evolution at the *Adh* locus in *Drosophila*. *Nature* 351: 652–654.

Nachman M. W. and Crowell S. L., 2000. Contrasting evolutionary histories of two introns of the Duchenne muscular dystrophy gene, *Dmd*, in humans. *Genetics* 155:1855–1864.

*Nielsen R., 2005. Molecular signatures of natural selection. *Annual Review of Genetics* 39: 197–218.

*Recommended reading

EXERCISES

9.1 A researcher compares a coding DNA sequence in mouse to the corresponding (homolog) sequence in humans and finds 420 nonsynonymous sites, 180 synonymous sites, 8 nonsynonymous mutations, and 6 synonymous mutations. What is the d_N/d_S ratio? Is there evidence for positive selection, negative selection, or no selection? (You do not have to do a statistical test—it is sufficient to provide a qualitative argument.)

9.2 In the following two tables, draw lines between each observation and the possible selective hypotheses that might explain the observation. Each observation may match more than one selective hypothesis, and each selective hypothesis may explain multiple observations.

Observation	Selective hypothesis
An increase in the proportion of low-frequency mutations	Negative selection acting on multiple mutations
An increase in the proportion of intermediate-frequency mutations	Positive selection acting on multiple mutations
A reduction in the number of segregating sites	
A d_N/d_S ratio > 1	
A d_N/d_S ratio < 1	Heterozygous advantage affecting a single mutation
An increase in the ratio of fixed to polymorphic sites	A recent selective sweep

9.3 A researcher has obtained the following counts of mutations by comparing DNA sequences from two different species ("between"), and by comparing a set of DNA sequences sampled from different individuals within a species ("within"):

	Within	Between
Nonsynonymous	12	24
Synonymous	16	8

Use these data to perform a MacDonald–Kreitman test. Which of the following factors might explain the result?
1. Positive selection
2. Negative selection
3. Balancing selection
4. Selective neutrality

9.4 Consider the following DNA sequence data obtained from a population:

```
Seq 1: atatacgatcgacagcctcgtctagtgctcgatatgccgc

Seq 2: acatacgatctacagcaccgtctagtgctcgatatgcagc

Seq 3: acatacgatcgacagcatcgtctagtgctcgatatgcagc

Seq 4: gcatacgatcgacagcctcgtcttgtgctcgatatgacgc

Seq 5: gcatacgatcgacagcctcgtctagtgctcgatatgaagc
```

Would the value of Tajima's D be positive or negative for these sequences? Does the value of Tajima's D suggest that either a selective sweep or balancing selection has affected the sequences, or is the result compatible with selective neutrality?

10 *Selection II: Interactions and Conflict*

IN CHAPTER 7, we introduced a simple type of natural selection that resulted from genotypic differences in the rate of survival from the zygote to adult stages. Limiting selection to this type allowed us to illustrate the basic principles of selection. However, there are many other ways that genes can affect survival and reproduction, and consequently many ways that allele frequencies can change because of selection. In this chapter, we will introduce several types of selection that are important in evolutionary biology and that can be understood by using basic principles of population genetics. The common theme will be selection that arises from interactions—either interactions between individuals in a population or interactions within the genome. We will not try to be comprehensive. Instead, we will show in a few cases how selection resulting from interactions can be analyzed by accounting for all the factors that affect survival and reproduction.

Selection on Sex Ratio

In many species, including all mammals and birds, there are separate sexes, males and females. In most such species, roughly equal numbers of males and females are born. For example, in humans, 1.05 males are born for every female. A long-standing evolutionary question is whether sex ratio at birth is an evolved trait in each species or a fixed property of each species that cannot evolve. R. A. Fisher was the first to ask how natural selection could alter the sex ratio and the first to show that the 1:1 sex ratio usually found is the expected outcome of selection. To analyze the evolution of sex ratios, we will introduce a method of analysis that is useful for answering a variety of other evolutionary questions.

Suppose that the probability that a newborn is a female is f and the probability that it is male is m ($m + f = 1$). For example, in a population of mice, suppose that 60% of the newborns are male, meaning that $m = 0.6$ and $f = 0.4$. Further, assume that there is no difference in the survival rates of males and females from birth to reproductive age. When mates are chosen, there are Nm males and Nf females, where N is the number of adults. In our example, if $N = 1000$, there will be 600 males and 400 females at the time of mating. If mating is random, each male has an equal chance of being the father of a newborn and each female has an equal chance of being the mother. Hence, the probability that a given male is a newborn's father is $1/(Nm)$, and the probability that a given female is the newborn's mother is $1/(Nf)$. Continuing with our example, the chance that each male is the father of a newborn is $1/600 \approx 0.00167$, and the chance that each female is the mother is $1/400 = 0.0025$.

The question is whether natural selection will tend to change m. To find an answer, we introduce an approach that is often used when analyzing complex selection models. We start by assuming that everyone in the population has the same m except for one individual who is heterozygous for an allele that causes m to be slightly different in its offspring: $m' = m + \delta m$. Then we ask whether this allele will tend to increase in frequency. If it does, we conclude that the sex ratio in the starting population is **evolutionarily unstable**. Mutations that alter the sex ratio will tend to increase in frequency, and the population's sex ratio will change. If instead we find that mutations that increase or decrease m do not tend to increase in frequency, then we conclude that the initial sex ratio is **evolutionarily stable**. In that case, the sex ratio will tend to remain the same.

When a sex ratio is evolutionarily stable, we say that it is an **evolutionarily stable strategy** (**ESS**). The term "strategy" suggests that each individual consciously chooses the sex ratio of its offspring, but of course that is not what happens. The sex ratio is determined by biological factors whose net effect is summarized by m and f. The goal is to determine what equilibrium is achieved when natural selection has the opportunity to modify factors that affect the sex ratio. We have already used the idea of evolutionary stability when we discussed heterozygote advantage. Even without finding the equilibrium allele frequency, we could conclude that both alleles would be maintained in a population, because we could show that both alleles tend to increase in frequency when they are rare. Therefore, a population homozygous for either allele is evolutionarily unstable, because the other allele will increase in frequency when introduced by mutation. Although we know that genetic drift can result in the loss of an advantageous allele, we imagine that alleles modifying the sex ratio occur sufficiently often that continued evolution of the sex ratio is possible.

To find the ESS sex ratio, we ask what happens to a mutation that alters the probability of a newborn being a male slightly, from m to $m' = m + \delta m$.

We do this by finding the expected contribution that the offspring make to the following generation. A fraction m' of the mutant individual's offspring will be male and each of those offspring contributes $1/(Nm)$ to the following generation. This assumes that the mutation is so rare that the reproductive success of the mutant individual's offspring is determined by the sex ratio of the nonmutant individuals. A fraction $1 - m'$ of the mutant individual's offspring will be female, and each contributes $1/(Nf)$ to the following generation. Therefore the average contribution of a mutant individual's offspring to the following generation is

$$\frac{m'}{Nm} + \frac{f'}{Nf} = \frac{m+\delta m}{Nm} + \frac{1-m-\delta m}{N(1-m)} = \frac{m}{Nm} + \frac{1-m}{N(1-m)} + \frac{\delta m}{N}\left(\frac{1}{m} - \frac{1}{1-m}\right) \quad (10.1)$$

The last term is the net difference between the contribution of the individual carrying the mutation to the following generation and one not carrying the mutation.

Suppose first that there is an excess of females ($m < 1 - m$). In that case, the coefficient of δm is positive. Therefore, if a mutation increases the proportion of males ($\delta m > 0$), individuals carrying the mutation will contribute more offspring to succeeding generations because the last term in Equation 10.1 is positive. Therefore, having more females than males ($m < \frac{1}{2}$) is not evolutionarily stable. The same argument applies if there is an excess of males ($m > 1 - m$). In that case, the coefficient of δm is negative, which tells us that a mutation that decreases the proportion of males ($\delta m < 0$) has an advantage. In our example above, we can see this result intuitively. If there are 600 males and only 400 females when mating takes place, each male is less likely than each female to be the parent of a newborn, because each newborn has only one father and one mother. Only if the numbers of males and females are equal will a mutation increasing or decreasing the sex ratio not increase in frequency. Therefore, we conclude that a 1:1 sex ratio is evolutionarily stable.

There is a nice experimental demonstration that a population will evolve to a 1:1 sex ratio if the sex ratio is artificially modified. In a population of the fruit fly *Drosophila mediopunctata*, the proportion of males at birth (m) was experimentally reduced to only 16% by fixing an allele at a locus on the X chromosome that alters the sex ratio. As discussed later in the chapter in the section on Meiotic Drive, some alleles increase or decrease the proportion of chromosomes that carry them. When the X chromosome has such an allele, the result is change in the sex ratio. After 49 generations of random mating in this population, m increased from 16% to 32% (**Figure 10.1**).

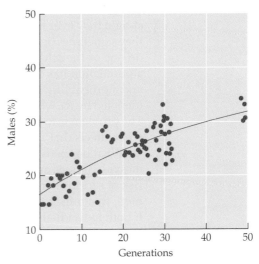

Figure 10.1 Evolution of the sex ratio in an experimental population of *Drosophila mediopunctata*. (After Carvalho et al., 1998.)

There have been many refinements to this line of argument that take account of the complexities of real species. Some generalizations are explored in the problems. But the basic approach is the same. You can determine whether a parameter is at an ESS by asking whether an allele that modifies it slightly will increase in frequency.

Resolving Conflicts

Part of Darwin's theory of natural selection is the struggle for existence that each organism engages in. Some of that struggle is against members of the same species for food, water, and other limited resources. Some of the competition is indirect: whoever gets to a resource first uses it and leaves less or none for others. Sometimes, however, there is direct competition: two or more individuals want the same thing at the same time. Most animals have the capacity to fight, and sometimes they engage in real fights that result in serious injuries or death. At other times, though, individuals do not have real fights, but instead engage in ritualized fighting by displaying their vigor in some way other than actual physical combat. A display may look ferocious, but it does not carry as much risk of injury to either participant as an actual fight. Such displaying or ritualized fighting is found in many animal species. For example, male red deer (**Figure 10.2**) bellow and display their antlers in their efforts to drive off other males. From the human perspective, each male seems to be trying to convince other males that he will win if they attempt to fight. Although deer have the capacity to seriously harm one another by kicking or by goring one another with their antlers, they rarely do so.

The observation that animals sometimes fight in earnest and sometimes engage in a ritualized display poses several evolutionary questions. Why should they ever fight in earnest? Or why should they ever display? And if they sometimes fight and sometimes display, how often should an individual do one or the other? To be more specific, suppose that each individual in a population has a probability p of engaging in a real fight (denoted by F) and a probability $1 - p$ of displaying, that is, engaging in a ritualized fight (denoted by R), when in a contest with another member of the same species. The problem is to find whether there is an ESS value of p and to determine what affects that value.

What happens to an individual in a contest depends both on what that individual does and on what the competitor does. If both fight, there is a risk of serious harm to both individuals. Furthermore, the risk is likely to more than offset the potential benefit of whatever is gained by winning. If both individuals display, there is much less risk to either of them. They both have a chance of gaining the contested item and therefore both have a chance of obtaining some net benefit. The cost of the ritualized fight is less than the gain from whatever it is they both want. If one fights and the other displays, then the one who starts to fight in earnest will win against

Figure 10.2 Red deer locking antlers in a fight. Many contests in red deer are resolved without actual fighting.

the other who was not planning to fight. The essence of the problem is that fighting results in a net benefit when the other displays but not when the other fights.

We can represent this kind of contest by a 2×2 matrix shown in **Table 10.1**. F and R are strategies adopted by each male, and $-b, c,$ and d indicate the payoff when each pair of strategies is adopted. In this simple example, $b, c,$ and d represent the change in an individual's viability as a result of the outcome of the dispute. How the viability changes depends on what is being disputed. If both individuals want the same food item, then failing to get it would increase the risk of starvation and getting it would increase the chance of survival. In many real situations, contests are over breeding sites or access to females, both of which affect success in mating and require a similar but more complicated analysis.

Based on our intuition about each outcome, b should be positive, which means that when both fight, the viability of both is reduced on average

TABLE 10.1   The costs and benefits to Individual 1 when it contests with Individual 2

		Individual 2	
		F	R
Individual 1	F	$-b$	c
	R	0	d

Note: F indicates fighting and R indicates ritualized displaying; $-b$ is the loss to 1 if both fight; c is the gain to 1 if it fights and 2 displays; d is the gain to 1 if both display. There is no gain to 1 if it displays and 2 fights.

because the risk of injury outweighs the benefit from obtaining what is being fought over. Similarly, d should be positive. On average, each has a 50% chance of winning a ritualized fight and so each should enjoy the benefit half the time. Whoever loses suffers no cost other than the effort required for displaying. When one fights and the other does not, the fighter wins every time, but the other suffers no loss because it simply withdraws from the contest. It is reasonable to assume that $c > d$, because there is some cost to displaying that is not borne by individuals who win because their competitor chooses not to fight.

Although our model of conflict is quite simple, it contains elements that are common to many situations involving conflict. It is similar to the classic prisoner's dilemma (**Box 10.1**). We can see that if everyone always displayed and never fought ($p = 0$), then everyone would be better off, because no one would suffer from engaging in a real fight. Everyone would get a benefit d on average. But we can also see that a population in which

---

BOX 10.1   The Prisoner's Dilemma

The problem of fighting and displaying in an evolutionary context is equivalent to the prisoner's dilemma frequently posed in game theory. Suppose that two people have been arrested for committing a crime that they actually did commit together. They are each taken to separate rooms and questioned at the same time and out of each other's hearing. Assume that each knows that the police have no evidence of their guilt. If neither confesses, they will be released. If one of them confesses and the other does not, he will be given a 1-year sentence and the other will receive a 10-year sentence. But if both confess, they will both be given 5-year sentences. What should the two prisoners do? The problem is summarized in a payoff matrix that is similar to the one in Table 10.1, with not confessing being equivalent to displaying and confessing being equivalent to fighting:

		Prisoner 2	
		Not confess	Confess
Prisoner 1	Not confess	0	−10
	Confess	−10	−5

Although not confessing is better for both, each knows that a confession by the other will make his own prospects worse, a situation well understood by criminal investigators. The difference between this situation and the evolutionary problem is that, in the evolutionary context, the decision has to be made repeatedly, while in the prisoner's dilemma, as posed here, a prisoner faces the decision only once.

$p = 0$ is evolutionarily unstable. A mutation that caused carriers to always fight would increase in frequency because carriers of that mutation would win every contest. However, a population in which every individual always fought ($p = 1$) would also be evolutionarily unstable. A mutation that caused carriers to never fight would increase in frequency because non-fighters would never run the risk of injury from fighting, even though they would not win any contests.

To find the ESS value of p, we consider a population in which every individual fights with probability p and displays with probability $1 - p$, independently of what the other individual does. The assumption that p has a fixed value does not account for the obvious possibility that a competitor can alter its behavior in response to what it thinks the other will do. We will not allow for that here but note that more sophisticated models of conflict in which an individual's behavior depends on the opponent's initial behavior can be analyzed in the same way. The average gain to an individual in our model is

$$G_p = p^2(-b) + p(1 - p)c + (1 - p)p(0) + (1 - p)^2 d \qquad (10.2)$$

The first term is the probability that both individuals fight (p^2) multiplied by the gain ($-b$) when that happens. The second term is the probability that the individual fights (p) and his opponent displays ($1 - p$) multiplied by the gain to the fighter (c). The third term is the probability that the individual displays ($1 - p$) and the other fights (p) multiplied by the benefit to the displayer (0). The last term is the probability that both display [($1 - p)^2$] multiplied by the benefit when that happens (d).

We next ask what is the average gain to an individual who carries a mutation that gives him a slightly different probability of fighting, $p' = p + \delta p$. The others it encounters will fight with probability p and display with probability $1 - p$. Therefore its average benefit to the individual carrying the mutant allele will be

$$\begin{aligned} G_p &= (p + \delta p)p(-b) + (p + \delta p)(1 - p)c + (1 - p - \delta p)(1 - p)d \\ &= p^2(-b) + p(1 - f)c + (1 - p)d + \delta p[-pb + (1 - p)c - (1 - p)d] \end{aligned} \qquad (10.3)$$

We now have the same kind of expression we had in Equation 10.1 when finding the ESS sex ratio. If the term in square brackets is positive, then a mutation that increases p ($\delta p > 0$) will increase in frequency, and if the term in square brackets is negative, then a mutation that decreases p ($\delta p < 0$) will increase in frequency. Therefore, the ESS value of p is the value that makes that term 0:

$$-pb + (1 - p)c - (1 - p)d = 0 \qquad (10.4)$$

which implies that

$$\hat{p} = \frac{c - d}{c - d + b} \qquad (10.5)$$

is the ESS value of p. In this expression, we see that $0 < \hat{p} < 1$, provided that there is an average loss when both fight ($b > 0$) and that there is greater gain from fighting than displaying when the other individual displays ($c > d$).

Although our model does not attempt to account for the many complexities of real behavior during contests, it does contain the essential trade-off between fighting and displaying and it does lead to the conclusion that natural selection should achieve a balance between fighting and displaying, a balance that depends on the relative costs of fighting and the benefits of winning contests.

Kin Selection

Altruistic behaviors are behaviors that benefit others at some cost to the individual performing the behavior. Such behaviors are seen in most animals. For instance, many birds and mammals emit a loud cry when they see a potential predator, even though such an outburst may draw the attention of the predator and increase the risk that the caller will be killed. Many birds engage in mobbing of predators, attacking a much larger predator singly or in groups despite the apparent risks associated with being so close to the predator (**Figure 10.3**). It is difficult to be sure that a particular behavior is truly altruistic. It may be, for example, that a bird that participates in mobbing is actually at a lower risk of being preyed upon than one who tries to hide in safety. Only careful study of each case can determine the costs and benefits of each behavior. Nevertheless, the animal behavior literature con-

Figure 10.3 Example of mobbing behavior.

tains so many examples of apparently altruistic behaviors that real altruism must be common. Altruistic behavior poses an evolutionary puzzle. Why should an individual do something that reduces its own chance of survival and reproduction even if that behavior increases the chances of survival and reproduction of others?

One explanation for the evolution of altruistic behaviors is that altruism is preferentially directed to close relatives who share genes with the altruistic individual. Selection that depends on both the cost to an individual and the benefit to close relatives is called **kin selection**. This is another area of population genetics theory that is well developed. We will illustrate the idea in the simplest possible context.

Assume that a population contains no individuals who perform a certain kind of altruistic behavior, say, giving a warning call when a potential predator is seen. Then suppose that a single individual is heterozygous for a mutation that causes it to give an altruistic warning call. It reduces the caller's chance of survival by c, the viability cost of the altruistic behavior. Suppose also that a close relative benefits from the behavior and has its chance of survival increased by an amount b, the benefit of the altruism. Because of the relationship between the two individuals, the one who benefits may also carry the mutation. The probability that two individuals carry the same allele inherited from a recent ancestor is the coefficient of relatedness, R, which is the average fraction of alleles shared by close relatives. For full siblings, $R = \frac{1}{2}$, for half-siblings, $R = \frac{1}{4}$, and so forth.

It is helpful to think about this problem from the mutation's point of view. This mutation will tend to increase in frequency if the increased chance that a relative survives offsets the decreased chance that the altruist survives. The lower viability of the altruist reduces the mutant frequency by c and the higher viability of the relative increases the mutant frequency by Rb. R must be included because that is the probability that the relative carries the altruistic mutation. The result then is that the mutation will tend to increase in frequency if

$$c < Rb \qquad (10.6)$$

This inequality is called **Hamilton's Rule**, first formulated by W. D. Hamilton.

It is important to remember that b represents the benefit specifically to a relative. If unrelated individuals also benefit from altruistic behavior, then the close relative has to benefit even more. With warning calls, for example, unrelated individuals may also have their chance of predation reduced when they hear the warning call, but for giving the warning call to be favored by kin selection, a close relative must have its chance reduced by some additional amount. The relative might be closer to the caller, for example, and hence be more likely to hear the call and find a safe refuge.

Hamilton's rule makes intuitive sense, but it is nonetheless based on a revolutionary idea. It changes our concern from the survival and reproduction of each individual to the survival and reproduction of an allele.

The fate of an allele depends, of course, on the survival and reproduction of individuals who carry it, but it is the collective chance of survival that determines whether an allele increases in frequency and not only the chance of each individual considered separately.

Insects in the order Hymenoptera, which includes bees, ants, and wasps, provide an interesting test of kin selection theory. Females in many species engage in an extreme form of altruism. They completely give up the ability to have offspring and instead develop into sterile workers who devote themselves to raising the offspring of one of their sisters (the queen). Males in these species do not help. They exist only to mate with future queens. This extreme asymmetry in the activities of the two sexes is accounted for by the unusual mechanism of sex determination in Hymenoptera. Females develop from fertilized eggs and are diploid. Males develop from unfertilized eggs and are haploid. Consequently, sisters share, on average, ¾ of their genomes. All of the paternal contribution is the same in each sister because the father is haploid and produces gametes that all have his complete genome. Half of the maternal contribution is the same in each sister because the mother is diploid and produces gametes that carry on average half of the mother's genome. In contrast, males share only ¼ of the genes of their sisters. They do not have a father and inherit only half of the mother's genome. This asymmetry in relatedness accounts for the difference in altruistic behavior in the two sexes. A confirmation of this explanation is provided by social termites, in which both males and females are diploid. In social termites there is no asymmetry: sterile workers are both male and female.

Kin selection is not the only explanation that has been proposed to account for altruistic behaviors. An alternative is **group selection**, which posits that altruistic behaviors evolve and persist because they benefit the species as a whole even if they are detrimental to the altruistic individuals. Group selection is appealing because it is in accord with what we would like be true in human populations. Because altruism is good for society as a whole, it should persist. Unfortunately, group selection as an evolutionary explanation is not supported by population genetic analysis. A population consisting only of individuals who are altruistic to everyone is not evolutionarily stable. An individual carrying a mutation that causes it to *not* be altruistic will benefit from the altruism of others but not bear the cost of being altruistic. Therefore its chance of surviving will be higher than the population average and the frequency of the non-altruistic allele will increase.

In order for altruism to be evolutionarily stable, there has to be some cost to non-altruistic individuals. In the case of kin selection, the cost to non-altruistic individuals is the reduced chance of survival of their relatives. An alternative is some type of social penalty, which is possible in species that have the capacity for individual recognition. If non-altruistic individuals are denied the benefits of altruistic behaviors by others, that could impose sufficient cost to make altruistic behavior evolutionarily stable. For example, an individual who fails to share its food with others may be left out of later food sharing.

Selfish Genes

Meiotic Drive

All of the evolutionary processes we have considered so far have assumed that Mendel's first law applies: heterozygous individuals produce equal numbers of gametes carrying each allele. That assumption is true for the vast majority of genes, but there are exceptions. A few genes carry alleles that cause heterozygotes to produce an excess of gametes carrying those alleles, a process called **meiotic drive**. For example, in domestic mice (*Mus musculus*), roughly 90% of the sperm produced by males heterozygous for the *t*-haplotype on chromosome 17 carry the *t*-haplotype. The mechanism causing this unequal segregation is not completely understood, but it is known that heterozygous males produce equal numbers of sperm of both types and then sperm carrying the normal haplotype become functionally inactivated.

The *t*-haplotype is an example of a selfish genetic element. It replicates itself disproportionally and, unless it is opposed by other forces, will reach fixation. In the case of the *t*-haplotype, fixation is prevented by the fact that males homozygous for *t* are sterile. In order to illustrate the way that selfish genetic elements evolve, we will consider a simple model that does not have the complexity created by having meiotic drive expressed only in one sex. Assume that at a locus with two alleles, *A* and *a*, one of the alleles is the driving allele: of the gametes produced by *Aa* heterozygotes, $(1 + r)/2$ are *A*-bearing and $(1 - r)/2$ are *a*-bearing. The parameter *r* is the strength of meiotic drive. If *A* and *a* are otherwise equal in their effects on survival and reproduction, then we can calculate the change in the frequency of *A* in one generation of random mating. We cannot assume Hardy–Weinberg genotype frequencies because Mendel's first law, on which the HW frequencies are based, no longer applies. But we can proceed in the same way that we did when deriving the Hardy–Weinberg frequencies and consider what happens in each family. **Table 10.2** shows the relevant details for the nine kinds of families. We find the genotype frequencies in the next generation by adding over families, taking account of the frequency of each family:

$$f'_{AA} = f^2_{AA} + (1+r)f_{AA}f_{Aa} + \frac{(1+r)^2 f^2_{Aa}}{4} = \left[f_{AA} + (1+r)f_{Aa}/2\right]^2$$

$$f'_{Aa} = 2f_{AA}f_{aa} + (1+r)f_{AA}f_{Aa} + f^2_{Aa}(1+r)(1-r)/2 + (1-r)f_{Aa}f_{aa}$$
$$= 2\left[f_{AA} + (1+r)f_{Aa}/2\right]\left[f_{aa} + (1-r)f_{Aa}/2\right] \qquad (10.7)$$

$$f'_{aa} = f^2_{aa} + (1-r)f_{Aa}f_{aa} + \frac{(1-r)^2 f^2_{Aa}}{4} = \left[f_{aa} + (1-r)f_{Aa}/2\right]^2$$

TABLE 10.2 Families and their offspring frequencies when there is meiotic drive

Mother	Father	Frequency	Offspring		
			AA	Aa	aa
AA	AA	f^2_{AA}	1	0	0
AA	Aa	$f_{AA}f_{Aa}$	$(1+r)/2$	$(1-r)/2$	0
AA	aa	$f_{AA}f_{aa}$	0	1	0
Aa	AA	$f_{Aa}f_{AA}$	$(1+r)/2$	$(1-r)/2$	0
Aa	Aa	f^2_{Aa}	$[(1+r)/2]^2$	$[(1+r)/2][(1-r)/2]$	$[(1-r)/2]^2$
Aa	aa	$f_{Aa}f_{aa}$	0	$(1+r)/2$	$(1-r)/2$
aa	AA	$f_{aa}f_{AA}$	0	1	0
aa	Aa	$f_{aa}f_{Aa}$	0	$(1+r)/2$	$(1-r)/2$
aa	aa	f^2_{aa}	0	0	1

Although the algebra is somewhat complicated, the result is simple and makes intuitive sense. The genotype frequencies in the next generation are the frequencies you would expect if you assume that two gametes are drawn randomly from a pool of gametes in which the composition is biased because of meiotic drive. The frequencies of A and a in the gamete pool is

$$f'_A = f'_{AA} + f'_{Aa}/2 = f_{AA} + (1+r)f_{Aa}/2 = f_A + rf_{Aa}/2 \quad (10.8)$$

Furthermore, the genotype frequencies in the next generation are the Hardy–Weinberg frequencies for f'_A and f'_a, the frequency of A in the next generation.

Even though meiotic drive alters the frequencies of gametes produced by the heterozygotes, it does so in a way that preserves the Hardy–Weinberg genotype frequencies, but with the allele frequencies changed. That is very convenient, because it allows us to see that meiotic drive of strength r is equivalent to additive selection with selection coefficient s when r is small in absolute value. In Equation 10.8, $f_{Aa} \approx 2f_Af_a$ when r is small. The result is equivalent to the equation in Box 7.4 for the change in allele frequency under additive selection when s is small.

Selection of intensity s causes a substantial change in allele frequency in $1/s$ generations, so the same will be true for meiotic drive of intensity r. This approximate argument is confirmed by the numerical iteration of Equation 10.7, as shown in **Figure 10.4**. Even very weak meiotic drive causes the driving allele to be fixed relatively quickly. Strong meiotic drive causes fixation in only a few generations (Figure 10.4).

Alleles causing meiotic drive may also directly affect survival and reproduction. In the t-haplotype system, for example, males homozygous

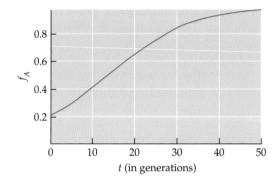

Figure 10.4 Prediction of Equation 10.7 for $r = 0.1$ and $fA(0) = 0.2$.

for t either fail to develop or are sterile. Obviously, selection of that type prevents fixation of the t-haplotype. In that case, a balance is reached between meiotic drive and selection against the driving allele.

Transposons

Transposons are short genomic regions that can move to other positions in the genome. There are many types of transposons and many mechanisms they use to insert themselves in new positions. Some types of transposons simply move from one place to another in the genome, excising themselves from one location and inserting themselves into the new location. Such transposons, which were first discovered in maize, do not count as selfish DNA, because they do not increase the number of copies. Instead, they change their map locations in a way that was quite unexpected when they were discovered by B. McClintock. Other types of transposons increase their copy numbers by duplicating themselves and inserting the new copy somewhere else in the genome without deleting the initial copy. A typical transposon of this kind is a **retrotransposon**, which is a transposon descended from an RNA retrovirus. The *copia* retrotransposon found in *Drosophila melanogaster* is one example. *Copia*, like all retroviruses, encodes a reverse transcriptase that allows the mRNA copy of *copia* to be reverse-transcribed into another location. The result of transposition of *copia* is an increase in copy number. This is another example of selfish DNA. The number of copies tends to increase because of the mechanism of replication and not because of any effect on survival and reproduction each transposon has.

To predict how the average number of copies of *copia* in a population of *D. melanogaster* changes with time, let i_j be the number of copies of *copia* in the jth individual in a population of size N. By summing over all individuals, we can find the average number of copies per individual,

$$\bar{i} = \frac{1}{N} \sum_{j=1}^{N} i_j \qquad (10.9)$$

Now consider the mating of two individuals, one with i_j copies and the other with $i_{j'}$ copies. In each individual, each copy has a probability λ of transposing and inserting the new copy somewhere else in the genome. Therefore, the individual with i_j copies will increase the number to $(1 + \lambda)i_j$ before meiosis. New copies of *copia* insert randomly in the genome. Therefore, when gametes are produced, each contains on average $(1 + \lambda)i_j/2$ copies. In saying this, we ignore the complication caused by the X chromosome. Similarly, the gametes of the other individual contain on average $(1 + \lambda)i_{j'}/2$ copies. Their offspring, then, have an average of

$$(1+\lambda)\frac{i_j+i_{j'}}{2} \tag{10.10}$$

copies. Now we take the average by summing over all pairs of parents to find

$$\bar{i}(t+1) = (1+\lambda)\bar{i}(t) \tag{10.11}$$

In other words, the average number of copies in the population will increase by a factor $(1 + \lambda)$ per generation.

As with meiotic drive, the basic biology of *copia* and similar transposons causes the number of copies to increase. Transposons are quite different, though, because they affect not just one locus but potentially the whole genome. What limits the number of copies of transposable elements depends on other factors. For *copia* and similar elements in *D. malanogaster*, it appears that they reduce fertility in part by causing recombination between copies in nonhomologous genomic regions, thus creating problems during meiosis. The number of copies of *copia* usually does not exceed fifty per genome of *D. melanogaster*. Other types of transposons are not so restricted in copy number, and a few have become so common that they constitute a substantial fraction of the genomes of higher animals. For example, roughly 17% of the human genome is made up of **LINEs (Long Interspersed Nuclear Elements)**, another type of transposon. In humans, transposons of many different types comprise nearly 50% of the total genome, and the vast majority, if not all, appear to serve no useful purpose. They are abundant because of their intrinsic tendency to replicate themselves in spite of any effect they have on survival and reproduction.

Species Formation

Groups of organisms are classified into different species if they are reproductively isolated from each other, meaning that they do not freely interbreed and produce viable and fertile offspring under natural conditions. There are many mechanisms of reproductive isolation. Some, including differences in karyotype, prevent the development of hybrid zygotes or cause hybrid adults to be sterile. For example, horses have 64 chromosomes and donkeys have 62. A male donkey mated to a female horse produces a mule, a vigorous and useful animal that is almost certain to be sterile. Other mechanisms prevent hybrid zygotes from forming by preventing mating of members of different species. Differences in the timing and location of mating prevent members of different species from encountering one another. Differences in mating calls of the males of many species of birds and insects cause reproductive isolation because females respond only to the calls of their own species.

One question that arises is what causes reproductive isolation to evolve. Darwin's answer was that populations geographically isolated from each other will evolve phenotypic differences, some of which will cause them to be reproductively isolated. But we can also ask whether reproductive isolation will evolve even if populations are not completely isolated geographically. One way this could happen is if selection favors different adaptations in different habitats. Suppose, for example, that a species has a mainland population and an island population and that the mainland population sends migrants to the island every generation. Suppose also that there is a phenotype on the island that enjoys high viability and that the immigrants from the mainland are of another, less favored, phenotype. That is the situation with the light- and dark-colored beach mice discussed at the beginning of Chapter 7. The lighter forms have higher viabilities on the sandy islands and the darker forms have higher viabilities on the mainland. We are supposing that, on the islands, continued immigration from the mainland reintroduces less-favored types and prevents the island population from becoming better adapted to its habitat. If individuals born on the island were to mate preferentially with others born on the island, the effect of immigration would be reduced. It seems reasonable, then, that selection would favor alleles in the island population that would help them discriminate against immigrants. The result would be assortative mating (mating with similar types) on the island. Complete assortative mating is equivalent to reproductive isolation of the mainland and island populations, in which case the two populations are actually different species. This process is called **reinforcement** because selection favoring assortative mating reinforces existing differences between the island and mainland population.

To understand reinforcement, we use a simple model. Assume there is a locus with two alleles, A and a, and that a is fixed on the mainland. On the island, A is slightly favored because it confers an advantage to conditions on the island that are not found on the mainland. Selection on $Mc1r$ in the beach mouse is an example of such selection. Selection on the island would result in the fixation of A, but that is prevented by immigration of a-bearing individuals from the mainland.

We first find what balance is attained between immigration of a and selection favoring A. The problem is similar to finding the balance between mutation to deleterious alleles and selection to remove them. We assume genic selection in favor of A, because that case is the easiest to analyze. We already found in Chapter 7 that for genic selection,

$$f_A' = \frac{f_A}{1-sf_a} \tag{10.12}$$

(Box 7.5). We also showed in Box 7.5 that the genotype frequencies in the adults are in their Hardy–Weinberg proportions with modified allele frequency:

$$f_{AA}' = f_A'^2; \; f_{Aa}' = 2f_A'f_a'; \; f_{aa}' = f_a'^2 \tag{10.13}$$

Now assume that adults from the mainland replace a fraction m of the adults on the island. On the mainland a is fixed so all the immigrants are aa. Therefore, after migration to the island,

$$f_{AA}'' = (1-m)f_{AA}'; f_{Aa}'' = (1-m)f_{Aa}'; \; f_{aa}'' = (1-m)f_{aa}' + m \tag{10.14}$$

As a consequence, the frequency of A after migration is $f_A'' = (1 - m)f_A'$. Random mating does not change that frequency, so f_A in the next generation is f_A''. There is an equilibrium frequency when f_A satisfies the equation

$$f_A = (1-m)\frac{f_A}{1-sf_a} \tag{10.15}$$

Solving for f_A, we find that

$$\hat{f}_A = 1 - m/s \tag{10.16}$$

is the equilibrium frequency on the island. For this to be a reasonable result, m has to be less than s. If $m > s$, immigration overpowers natural selection and prevents A from being maintained on the island at all. Note that we have not had to assume that either m or s is small. This result is true even if there is very strong selection on the island opposed by substantial immigration from the mainland.

Now suppose that there is a second locus with alleles B and b, unlinked to the A/a locus, that affects mating preference. Assume bb individuals make no distinction between residents and immigrants when they chose mates, but Bb and BB individuals discriminate perfectly against immigrants. They can by some means distinguish individuals born on the island and mate only with them. In this simple model, B is a dominant allele that causes complete assortative mating.

Suppose that A is at its equilibrium frequency and B is initially in a low frequency f_B. Also suppose that the two loci are in linkage equilibrium (LE). The assumption of LE is approximately correct and greatly simplifies the analysis. Now consider the average viability of offspring of an average newborn *bb* individual. The frequency of *A* is f_A, so the average viability is

$$\bar{v}_{bb} = f_A^2 + 2(1-s)f_A f_a + (1-s)^2 f_a^2 = \left[1 - s(1 - f_A)\right]^2 \qquad (10.17)$$

where f_A is given by Equation 10.15. The average viability of a newborn *Bb* individual, $\bar{v}_{Bb}$, will differ from $\bar{v}_{bb}$ because its parents did not mate with immigrants. The frequency of *A* among newborn *Bb* individuals is given by Equation 10.12, and hence

$$\bar{v}_{Bb} = f_A'^2 + 2(1-s)f_A' f_a' + (1-s)^2 f_a'^2 = \left[1 - sf_a'\right]^2 \qquad (10.18)$$

Therefore, the ratio of these average viabilities tells us the advantage *B*-bearing individuals have when *B* is in low frequency:

$$\frac{\bar{v}_{Bb}}{\bar{v}_{bb}} = \left[\frac{1 - sf_a'}{1 - sf_a}\right]^2 \qquad (10.19)$$

We know that $f_a' < f_a$, because *a* is selected against on the island, and hence the ratio in Equation 10.19 is greater than 1. Therefore we conclude that mutations that cause reproductive isolation on the island are favored by natural selection.

We can determine the strength of selection in favor of *B* by rewriting Equation 10.19 as

$$\frac{\bar{v}_{Bb}}{\bar{v}_{bb}} = \left[\frac{1 - sf_a + s(f_a - f_a')}{1 - sf_a}\right] = \left[1 + \frac{s(f_a - f_a')}{1 - sf_a}\right]^2 \qquad (10.20)$$

If *s* and *m* are not too large, the advantage to *B* is proportional to *s* multiplied by the increase in the frequency of *a* caused by immigration, which is proportional to *m*. Therefore, even though we have assumed that a single mutation, *B*, creates reproductive isolation in one step, we find that selection in favor of *B* is weak unless both *s* and *m* are large.

References

*Coyne J. and Orr H. A., 2004. *Speciation*. Sinauer Associates, Sunderland, MA.

*Dawkins R., 1976. *The Selfish Gene*. Oxford University Press.

Maynard Smith, J., 1964. Group selection and kin selection. *Nature* 201: 1145–1147. http://www.nature.com/nature/journal/v201/n4924/abs/2011145a0.html

*Maynard Smith, J., 1982. *Evolution and the Theory of Games*. Cambridge University Press.

Orgel L. E. and Crick F. H. C., 1980. Selfish DNA: The ultimate parasite. *Nature* 284: 604–607. http://www.nature.com/nature/journal/v284/n5757/abs/284604a0.html

*Recommended reading

EXERCISES

10.1 Assume that a species produces 40% males and 60% females at birth. What is the selection coefficient for a rare mutant that increases the proportion of males to 45% in heterozygous carriers?

10.2 In many species, males and females are of different sizes and require unequal investment on the part of the parents. Suppose that a male requires twice as much investment as a female. In that case, a mutant that increases the proportion of males from m to $m' = m + \delta m$ will decrease the proportion of females by twice as much, $f' = f - 2\delta m$. What do you predict the ESS sex ratio will be for such a species? (Hint: Start with the first term in Equation 10.1.)

10.3 Male red deer compete for access to females during mating season. Males either display by bellowing and showing off their antlers or they fight by locking antlers or kicking one another. There is a real risk of injury and death in fights. There is some cost to engaging in long displaying contests because of the energy used and the lost time during which the males could have been searching for other females. Suppose you have carried out a long field study in which you have estimated the costs and benefits of fighting and summarized your results in the payoff matrix below:

	Male encountered	
Focal male	Fight	Display
Fight	−10	+8
Display	0	+5

   a. What would you predict the ESS frequency of displaying to be?

   b. Suppose that for some reason, there are fewer females to fight over. The main result is that the cost to prolonged displays increases because, if the a male waits too long, there may be no females left to mate with. Will selection favor an increase or decrease in p if the population was at the equilibrium established for before the number of females declined?

10.4 Suppose that a population of mice live on a rock which has small holes that provide shelter from predators. A mouse in a hole has a 5% chance of being eaten by a predator and a mouse not in a hole has a 25% chance of being eaten. You study these mice and find that when two mice want the same hole, half of the time they fight and risk serious injury from vicious biting and half of the time they merely display their little teeth and risk no injury at all. If this population is at an ESS for fighting frequency, what can you conclude about the mortality rate when two mice fight?

10.5 Well before the theory of kin selection was proposed and worked out, the famous evolutionary biologist J. B. S. Haldane was asked whether he would give up his life for his brother. He answered that he would not for one brother but he would for two brothers or eight cousins. What do you think he meant?

10.6 Prairie dogs commonly give warning calls when a potential predator is seen. Field studies of prairie dogs have shown that individuals tend to give warning calls when their close relatives are near enough to hear the call and tend not to do so otherwise. Are these observations consistent with the theory of kin selection or the theory of group selection?

10.7 Segregation distortion is a form of meiotic drive that occurs in *Drosophila melanogaster* when an X-linked locus carries an allele that causes the destruction of Y-bearing sperm. Suppose there are two alleles, D and d, and that D is the distorting allele: males hemizygous for D produce only X-bearing gametes.

a. Suppose that D is present in a population at 50% frequency. What will be the sex ratio in that population in the next generation? Assume the same mating success for D-bearing and d-bearing males. The answer does not depend on the sex ratio in the initial generation.

b. What will be the frequency of D in males and females in the next generation? (Hint: Assume random mating and compute the frequencies of all six types of families. Then determine the genotypes of the offspring of each family.)

c. What will happen to the population as D continues to increase in frequency?

10.8 Assume that an allele A increases in frequency because of a weak meiotic drive of strength $r = 0.01$. Use the theory of haploid selection (see Box 7.1) to find how long it would take A to increase in frequency from 0.1 to 0.10 under the effects of meiotic drive alone.

10.9 Assume that A in problem 10.8 is a recessive deleterious allele with selection coefficient 0.05 against it. What is the equilibrium frequency when the increase under meiotic drive is balanced by the reduction in frequency because of selection? (Hint: Use the formula from Box 7.4 that the change in f_A under selection is approximately when s is small.)

10.10 What is the selection intensity in favor of a rare allele that causes reproductive isolation in an island population if s on the island is 0.5 and the immigration rate is 0.1?

10.11 Suppose that a recessive allele has a selective advantage on an island that receives immigrants at rate m from a population fixed for the other allele. Let the relative viabilities on the island be 1 (AA), $1-s$ (Aa), and $1-s$ (aa) and assume that the immigrants all carry a. Assume $s = 0.5$ and $m = 0.1$. If $f_A = 0.01$, will A increase in frequency? (Hint: Assume Hardy–Weinberg genotype frequencies in the zygotes and then find the frequencies in the adults before immigration. Then change the frequencies because of immigration of a.)

11 *Quantitative Genetics*

IN EARLIER CHAPTERS, we were concerned with one or two loci at which alleles can be distinguished either by DNA sequencing or by their effect on phenotype. Unfortunately, most phenotypic characters of interest to evolutionary biologists and human geneticists do not have such a simple genetic basis. Many phenotypic traits, including height, weight, hair and eye color, blood pressure, and serum cholesterol levels, are affected by numerous genes. Traits such as these, which can be quantified on some scale, are called **quantitative characters**. We know that genes are important for quantitative characters because offspring resemble their parents with regard to these characters more than they do others in the same population, but the effect of individual genes on quantitative characters is not apparent.

Quantitative genetics is the study of the inheritance of quantitative characters. It is one of the oldest subjects in biology. Many of the methods now used in quantitative genetics, including some we will introduce in this chapter, grew from the practice of animal and plant breeding and were developed before the rediscovery of Mendelism in 1900. Yet quantitative genetics is a modern and rapidly developing subject, as well. Modern genomics is making it possible to map loci that affect quantitative characters, and modern methods of molecular biology are making it possible to understand how individual loci affect quantitative characters.

We will first present the biometrical theory of quantitative genetics based on phenotypic measurements. Next, we will relate the biometrical analysis to the properties of genes affecting the character of interest. Then we will describe how genomic methods are being used to map and characterize genes that affect quantitative characters.

Biometrical Analysis

Biometrical analysis, or **biometry**, is the study of quantitative characters in a population and the similarity of quantitative characters in close relatives. Many commonly used methods in statistics, including regression and correlation, had their origin in biometry. We consider a character that can be measured on a continuous scale, for example, height in humans. The first systematic study of height was published in 1889 by F. Galton. Some of his data for men are shown in **Figure 11.1**. This "bell-shaped" distribution is typical of most quantitative characters, and has a simple mathematical form described in **Box 11.1**. Values of x are concentrated near the maximum point, but some are farther away. The mean $\bar{x}$ and variance V of a bell-shaped distribution conveniently summarize its properties. In Figure 11.1, the mean, $\bar{x}$, is 68.5 in. and the variance, V_x, is 6.1 in². The standard deviation, $\sigma_x = \sqrt{V_x}$, is 2.47 in.

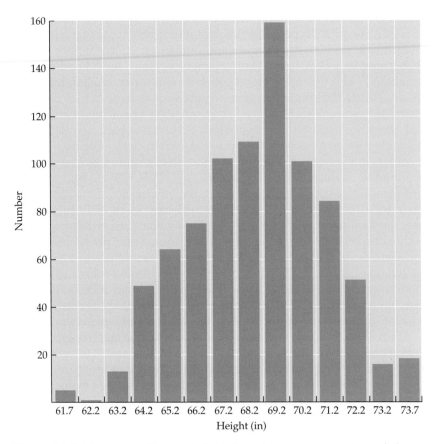

Figure 11.1 Histogram of human heights based on measurements made by F. Galton. The numbers of individuals in each height category are shown. (After Galton, 1889.)

BOX 11.1 Normal Distribution

The bell-shaped distribution of a quantitative character illustrated in Figure 11.1 is often seen when data are graphed. The probability that an individual has measurement x in a population depends on two parameters, the mean and the variance. The distribution has a simple mathematical form:

$$\Pr(x \mid \bar{x}, V) = \frac{1}{\sqrt{2\pi V}} e^{-(x-\bar{x})^2/(2V)}$$

where $\bar{x}$ is the mean and V is the variance. The interpretation of this distribution is as follows: the probability that x falls between any two values x_1 and x_2 is the area under the curve between x_1 and x_2:

$$\Pr(x_1 \leq x \leq x_2) = \frac{1}{\sqrt{2\pi V}} \int_{x_1}^{x_2} e^{-(x-\bar{x})^2/(2V)} \, dx$$

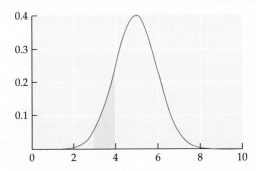

as illustrated in the figure on the top. The area of the shaded region is the probability that x is between 3 and 4. With $\bar{x} = 5$ and $V = 1$, that probability is about 0.136. The factor that multiplies the exponential is chosen so that $\Pr(-\infty < x < +\infty) = 1$.

This distribution is called a *normal* distribution because it appears in many contexts and the *Gaussian* distribution, after Carl Gauss who wrote extensively on the properties of this distribution. The parameters are easy to interpret: $\bar{x}$ is the value of x at which the curve has its maximal value and $\sigma = \sqrt{V}$ (the standard deviation) measures how broad or narrow the curve is.

The probability that x is between $\bar{x} - 1.96\sigma$ and $\bar{x} + 1.96\sigma$ is 0.95, as illustrated in the figure on the bottom. The reason a normal distribution arises so often is that whenever many random numbers are added, the sum tends to be normally distributed. That is called the **Central Limit Theorem**. It is no surprise then that quantitative characters are often normally distributed. They result from the interaction of many genes and many random events during development.

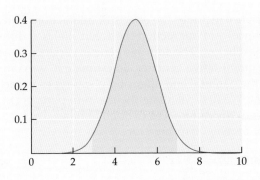

One question we can ask about data of the type shown in Figure 11.1 is how much of the variation is attributable to genetic differences between individuals and how much to environmental differences. Common experience tells us that both genes and environment contribute to a trait like height. We know that taller individuals tend to have taller offspring and we know that poor diet during adolescence stunts growth. To go beyond the obvious statement that both genes and environment contribute to a trait, we have to be able to quantify the contribution of each. To do so, we assume that a phenotypic measurement x is the sum of two factors:

$$x = g + e \tag{11.1}$$

where g represents the net contribution of all genetic factors to x and e represents the net contribution of all environmental factors. We assume that genetic and environmental contributions are independent of each other. In that case, the variance in x is the sum of two variances: V_G, which is the variance among individuals in all genetic factors that contribute to the trait in question, and V_E, the variance in all environmental factors. The ratio

$$h_B^2 = \frac{V_G}{V_G + V_E} \tag{11.2}$$

is called the **broad-sense heritability**. It is the fraction of the total variance in x that is attributable to genetic differences.

The genetic component, g, can be further divided into a part that is transmitted to each offspring and a part that is not transmitted. The part that is transmitted is called the **additive genetic component** and the rest is called the **interaction component**. The reason for making this distinction is that the contribution of some alleles to a trait depends on which other alleles are present. The genetic variance V_G is divided into the additive genetic variance (V_A) and the interaction variance (V_I):

$$V_G = V_A + V_I \tag{11.3}$$

Here, the interaction variance includes the effects of interactions between alleles at a locus (dominance) and interactions between alleles at different loci (epistasis), as discussed below. The additive genetic variance is of fundamental importance in quantitative genetics. To estimate it, we compute the covariance between the parental and offspring phenotypes in each set of parent–offspring pairs. We first calculate the mean in parents ($\bar{x}$) and offspring ($\bar{y}$), and then sum the differences over all families, $i = 1, \dots, n$, where n is the number of parent–offspring pairs:

$$Cov(x, y) = \frac{1}{n} \sum_{i=1}^{n} (x_i - \bar{x})(y_i - \bar{y}) \tag{11.4}$$

The covariance tells us the extent to which x_i and y_i tend to deviate from their mean values in the same way. If $Cov(x, y) > 0$, then we know that when $x_i > \bar{x}$, y_i tends to be greater than $\bar{y}$. If $Cov(x, y) < 0$, then the opposite is true;

TABLE 11.1 Galton's measurements of height

Midparent height	Offspring heights											
	62.2	63.2	64.2	65.2	66.2	67.2	68.2	69.2	70.2	71.2	72.2	73.2
72.5							1	2	1	2	7	2
71.5				1	3	4	3	5	10	4	9	2
70.5		1		1	1	3	12	18	14	7	4	3
69.5		1	16	4	17	27	20	33	25	20	11	4
68.5		7	11	16	25	31	34	48	21	18	4	3
67.5	3	5	14	15	36	38	28	38	19	11	4	
66.5	3	3	5	2	17	17	14	13	4			
65.5		9	5	7	11	11	7	7	5	2	1	
64.5	1	4	4	1	5	5		2				

Source: Galton (1889).

smaller-than-average x_i values are associated with larger-than-average y_i and vice versa.

The covariance between parents and offspring provides an estimate of V_A:

$$Cov(x,y) = \frac{V_A}{2} \qquad (11.5)$$

The factor of ½ is needed because each parent contributes only half of the genes in the offspring.

To illustrate, we consider the data on height in parents and offspring collected by Galton in the 1880s and presented in **Table 11.1**. In this study, Galton multiplied the heights of women by 1.08 to make them comparable to men. In this data set, Galton presented data for the mid-parent value, which is the average of the heights of the two parents. In **Box 11.2**, we show that the covariance between the mid-parent height and offspring height is expected to be the same as the covariance between that of either parent and the offspring. In Table 11.1, the covariance between the mid-parent and offspring heights is 1.6 in², so the estimate of V_A is 3.2 in².

The additive genetic variance is a fraction of the total variance. This ratio is the **narrow-sense heritability**, denoted by h^2:

$$h^2 = \frac{V_A}{V_x} \qquad (11.6)$$

For the data in Table 11.1, the variance of the mid-parent height is 2.77 in². The total variance in the population is twice the variance of the mid-parent height, 5.54 in² (see Box 11.2). Therefore, $h^2 = 3.2/5.54 = 0.58$ in Galton's data.

BOX 11.2 Variance of the Mid-parental Value

In empirical studies of heritability, when information from both parents is available, the mid-parent value of the quantitative character is used:

$$x_{MP} = \frac{x_M + x_F}{2}$$

where x_M is the character in the mother and x_F is the character in the father. The covariance between the mid-parental value and the offspring is

$$Cov(x_{MP}, x_O) = \frac{Cov(x_M, x_O) + Cov(x_F, x_O)}{2} = \frac{V_A}{2}$$

For most characters, the two covariances on the right-hand side are equal, in which case the covariance with the mid-parental value is the same as the covariance with either parent.

The variance of the mid-parental value differs, however:

$$Var\left(\frac{x_M + x_F}{2}\right) = \frac{Var(x_M)}{4} + \frac{Cov(x_M, x_F)}{2} + \frac{Var(x_F)}{4}$$

If mating is random with respect to the character, $Cov(x_M, x_F) = 0$. If the variances in males and females are equal, then

$$Var(x_{MP}) = \frac{Var(x_M)}{2} = \frac{Var(x_F)}{2} = \frac{Var(x)}{2}$$

The heritability of the character, $h^2 = V_A/V_x$, V_A is estimated by $2Cov(x_{MP}, x_O)$, and V_x is estimated by $2Var(x_{MP})$. The factors of 2 cancel, and the heritability is estimated by the ratio

$$h^2 = \frac{Cov(x_{MP}, x_O)}{Var(x_{MP})}$$

Heritability is important in quantitative genetics because it allows us to predict what happens when viability depends on the character. Suppose we study a character in a population with a known mean and variance. Then individuals are selected to form the parents of the next generation. Whether selection is done by a breeder (artificial selection) or is the result of the population's interactions with its environment (natural selection) does not matter. It matters only that an individual's phenotype determines to some extent whether it will survive to breed or not. As in our discussion of viability selection in Chapter 7, we can think of the adults who survive as a subset of the population of newborns. The mean of the character in the selected adults, $\bar{x}_s$, describes the net effect of selection. If $\bar{x}_s > \bar{x}$, then individuals with larger than average values of the character have a greater chance of surviving to breed. The **selection differential**, **S**, is the change in the mean caused by selection: $S = \bar{x}_s - \bar{x}$. A central result in quantitative genetics is the **breeder's equation**, which predicts the mean of the character in the offspring of the selected adults:

$$R = \bar{x}' - \bar{x} = h^2 S \tag{11.7}$$

where $\bar{x}$ is the mean in the offspring of the surviving parents.

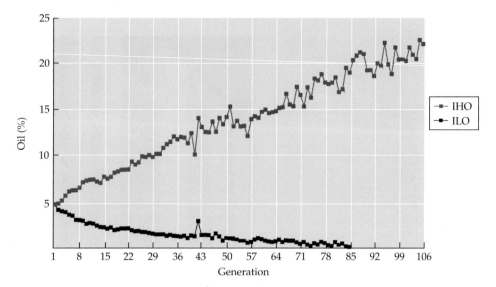

Figure 11.2 Results of continued selection for increased and decreased corn oil percentage since 1896. Each point represents the average oil content in corn kernels in each line each summer. Corn (*Zea mays*) is an annual species so there is only one generation per year. Selection was performed by selecting ears that had the highest or lowest 20% oil content. High Oil (IHO), and Low Oil (ILO). (After Dudley, 2007.)

Equation 11.7 can be used in two ways. If h^2 has been estimated, then 11.7 predicts the response of the character mean to selection. In fact, as long as h^2 does not change, it predicts the response for subsequent generations as well. In practice, the breeder's equation usually provides reasonably accurate predictions for at least 5–10 generations. After that, there may be a plateau at which the average remains at the same value for several generations.

One such study of selection has been on oil content in corn. Corn is an annual species, so each generation is the result of selection applied during the previous year. Populations were selected for higher and lower oil content. As shown in **Figure 11.2**, there is roughly a linear increase in oil content in the lines selected for higher values, which indicates that the heritability of oil content has not changed during this time. The line selected for lower oil content has reached the minimum value, and no further progress can be made.

A second use of the breeder's equation is to estimate the heritability from the response to selection. Heritability estimated by this method is called the **realized heritability**. **Figure 11.3A** shows the response to selection for ethanol vapor tolerance in *Drosophila melanogaster*. Selection on this trait is relatively easy to perform, thanks to a device called an inebriometer (**Figure 11.3B**), developed by K. E. Weber. The inebriometer measures tolerance of a fly to ethanol vapor by finding the point in a concentration gradient at which

(A)

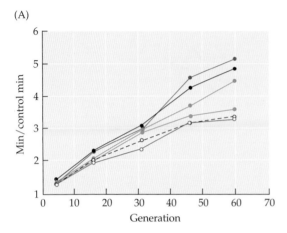

(B)

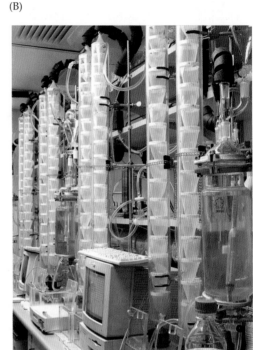

Figure 11.3 (A) Response to selection for ethanol tolerance in *D. melanogaster*. (B) Inebriometer, a device that measures the tolerance of individual flies to alcohol vapor. The inebriometer creates a gradient of alcohol concentration. Flies are attracted to alcohol because it is produced by fermenting yeast, which are an important source of food. They fly in the direction of increasing alcohol concentration until they reach their physiological limit of tolerance, at which time, they fall into different containers. In effect, the flies select themselves. By counting the flies in each container, the distribution of alcohol tolerance in the population is determined with relative ease. (After Weber and Diggins 1990; photo courtesy of Ulrike Heberlein.)

an individual can no longer fly. The realized heritability of vapor tolerance was about 0.22. For a character like ethanol tolerance in flies, it would be impractical to find the parent–offspring covariance for a large enough number of parent–offspring pairs to obtain an accurate estimate of the heritability.

Breeding Value

We can get an intuitive understanding of additive variance by considering the **breeding value** (bv) of an individual. We start by imagining that we can mate each male in a population to a large number of randomly chosen females and that each mating produces multiple offspring. Number the females $i = 1, 2, \ldots$ and let the mean of the offspring of male j mated to female i be y_{ij} and the average of the y_{ij} over i be m_j. The overall mean $\bar{x}$ is the average of m_j over all j. The breeding value of male j is twice the difference between the mean of its offspring and the overall mean:

$$bv_j = 2(m_j - \bar{x}) \tag{11.8}$$

TABLE 11.2 Example of estimated breeding values

	Male				
	A	**B**	**C**	**D**	**E**
	687	618	618	600	717
	691	680	687	657	658
	793	592	763	669	674
	675	683	747	606	611
	700	631	678	718	678
	753	691	737	693	788
	704	694	731	669	650
	717	732	603	648	690
Average	715.0	665.1	695.5	657.5	683.2
Deviation	31.74	−18.16	12.24	−25.76	−0.06
Breeding value	63.48	−36.32	24.48	−51.52	−0.12

Note: Five white leghorn roosters were each mated to eight females, and one male offspring from each mating was weighed at age eight weeks. Weights are in grams. The overall average is 683.26 g.
Source: Becker (1992).

The breeding value quantifies the net effect of genes transmitted from each male to its offspring and therefore is the additive effect described in the previous section. The factor of 2 is there because each male provides only half of the genes in each offspring. The other half come from the mother, who is chosen at random from the population.

In **Table 11.2**, we illustrate how to estimate the breeding value in chickens. Five roosters were each mated to eight hens and the weight at six weeks of one offspring from each family was recorded. The breeding values of the five males differ, with some being positive and some negative. Note that the average breeding value is necessarily zero because the breeding value is defined relative to the overall mean, $\bar{x}$.

So far, we have defined the breeding value for males in a species with separate sexes. However, a breeding value can be defined for every individual in a population, even when practical constraints prevent us from measuring it. Suppose we had the breeding values of every member of the population, $bv_1, bv_2, \ldots$. Because the mean breeding value is zero, the variance in breeding values is the average of the squared values for each individual:

$$V_A = \frac{1}{n}\sum_{j=1}^{n}\left(bv_j\right)^2$$

(11.9)

where n is the number of individuals. The variance in breeding values, denoted by V_A, is the additive genetic variance. In the example in Table 11.2, $V_A = 1720.49$ gm^2.

Quantitative Trait Loci

The concept of breeding value allows us to connect biometrical ideas to the effects of individual loci. If the genotype at a locus affects the average of a quantitative character, then that locus is called a **quantitative trait locus** (**QTL**). For example, a single nucleotide change in the 3′ untranslated region of the *HMGA2* locus in humans affects average height. Average heights of men and women with the three genotypes of this SNP are shown in **Table 11.3**. Although the effect of this SNP on height is small, it does contribute to the additive genetic variance. We can use the definition of the breeding value to determine how much a single QTL contributes.

Let x be the measurement of the quantitative character and assume that a locus with two alleles, A and a, affects x. The means of individuals with the three genotypes are x_{AA}, x_{Aa} and x_{aa}. For each genotype, assume that the character is normally distributed about these means with variance V_E, which we call the **environmental variance**. In this context, the environmental variance accounts for everything else that affects the character, including both random events during development and the genotypes of other loci. We illustrate these assumptions in **Figure 11.4**, which shows the distributions of x for the three genotypes, with averages 7, 8 and 9, and the overall distribution of x when $f_A = 0.4$. Although the genotype at this QTL affects the character on average, the measurement of x in any individual does not indicate the genotype of this locus. To relate this model of a single QTL to the breeding value and additive genetic variance, we start by assuming a randomly mating population in which A has frequency f_A. The mean of x in this population is

$$\bar{x} = f_A^2 x_{AA} + 2 f_A f_a x_{Aa} + f_a^2 x_{aa} \tag{11.10}$$

The variance of x is made up of two components, the component attributable to difference in genotype at the QTL and the component that is independent of the genotype at that locus. We denote the genetic component by V_G and compute it by finding the average of the squared deviations from the mean:

$$V_G = f_A^2 (x_{AA} - \bar{x})^2 + 2 f_A f_a (x_{Aa} - \bar{x})^2 + f_a^2 (x_{aa} - \bar{x})^2 \tag{11.11}$$

The total variance in x is

$$V_x = V_G + V_E \tag{11.12}$$

assuming that the environmental component is independent of the effect of the genotype at this QTL.

To compute the breeding value of individuals with each of the three genotypes, we start with **Table 11.4**, which shows the average of x in the offspring of each of the nine possible families.

TABLE 11.3 Data showing QTL affecting human height (cm)

	x_{TT}	x_{TC}	x_{CC}
Males ($n=3023$)	174.8	175.7	175.9
Females ($n=3508$)	161.9	162.4	162.5

Note: The average heights of males and females with each genotype at a SNP on chromosome 12 (denoted by rs1042725) are presented.
Source: Weedon et al. (2007).

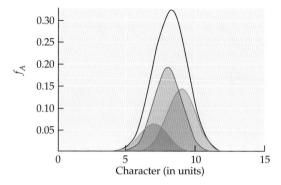

Figure 11.4 Illustration of the model of a QTL. The three filled curves show the distributions of the hypothetical character among individuals with each of the three genotypes. The means of the character are $x_{AA} = 7$, $x_{Aa} = 8$ and $x_{aa} = 9$ (in arbitrary units) and the variance of each distribution (V_E) is 1. The areas under each curve are adjusted to be the genotype frequencies when $f_A = 0.4$ (i.e., 0.16, 0.48, and 0.36). The black curve is the sum of the three filled curves. It shows the overall distribution of the character in this hypothetical population. Although the black curve is not a normal distribution because it is the sum of three normal distributions with different mean values, it would be difficult to distinguish it visually from a normal distribution.

From Table 11.4 and the Hardy–Weinberg frequencies, we compute the breeding values for the general one-locus model shown in **Table 11.5**.

An important special case of this model is the **additive model**, in which each copy of A increases the mean by the same amount, α, i.e., $x_{Aa} = x_{aa} + \alpha$ and $x_{AA} = x_{aa} + 2\alpha$. From Equations 11.10 and 11.11,

TABLE 11.4   Average offspring phenotype in each family

Mother	Father	Average offspring
AA	AA	x_{AA}
AA	Aa	$x_{AA}/2 + x_{Aa}/2$
AA	aa	x_{Aa}
Aa	AA	$x_{AA}/2 + x_{Aa}/2$
Aa	Aa	$x_{AA}/4 + x_{Aa}/2 + x_{aa}/4$
Aa	aa	$x_{Aa}/2 + x_{aa}/2$
aa	AA	x_{Aa}
aa	Aa	$x_{Aa}/2 + x_{aa}/2$
aa	aa	x_{aa}

Note: Genotypic means are x_{AA}, x_{Aa} and x_{aa} and Mendel's first law is assumed.

TABLE 11.5 Breeding values for general one-locus model

Parental genotype	Breeding value (bv)
AA	$2\left[f_A^2 x_{AA} + 2f_A f_a(x_{AA}/2 + x_{Aa}/2) + f_a^2 x_{Aa} - \bar{x}\right]$ $= 2\left[f_A f_a(x_{AA} - x_{Aa}) + f_a^2(x_{Aa} - x_{aa})\right]$
Aa	$2\left[f_A^2(x_{AA}/2 + x_{Aa}/2) + 2f_A f_a(x_{AA}/4 + x_{Aa}/2 + x_{aa}/4) + f_a^2(x_{Aa}/2 + x_{aa}/2) - \bar{x}\right]$ $= f_A^2(-x_{AA} + x_{Aa}) + f_A f_a(x_{AA} - 2x_{Aa} + x_{aa}) + f_a^2(x_{Aa} - x_{aa})$
aa	$2\left[f_A^2 x_{Aa} + 2f_A f_a(x_{Aa}/2 + x_{aa}/2) + f_a^2 x_{aa} - \bar{x}\right]$ $= 2\left[f_A^2(x_{Aa} - x_{AA}) + f_A f_a(-x_{Aa} + x_{aa})\right]$

Note: The breeding values are obtained from Table 11.3 by multiplying the probability of mating with an individual with each of the three genotypes, which are assumed to be in their Hardy-Weinberg proportions.

$$\bar{x} = x_{aa} + 2f_A \alpha \tag{11.13}$$

and

$$V_G = 2f_A f_a \alpha^2 \tag{11.14}$$

For the additive model, the breeding values reduce to simple expressions, shown in **Table 11.6**, which are obtained by substituting the expressions for x_{AA} and x_{Aa} into Table 11.5.

Because the average breeding value is 0, the additive genetic variance is the weighted average of the squared breeding values:

$$V_A = f^2_A[2f_a\alpha]^2 + 2f_A f_a[-\alpha(f_A - f_a)]^2 + f^2_a[-2\alpha f_A]^2 = 2f_A f_a \alpha^2 \tag{11.15}$$

Equations 11.14 and 11.15 shows that $V_G = V_A$ for the additive model. Therefore, all of the genetic variance is additive in this special case. Notice that V_A depends not only on α but also on the allele frequency. In a different population, a QTL may have the same effect on the character (α) but contribute more or less to V_A because f_A differs.

The general model is analyzed the same way, but the algebra is slightly more complicated. It is convenient to choose the parameters so that the difference from the additive model is apparent. Let $x_{AA} = x_{aa} + 2\alpha$ and $x_{Aa} = x_{aa} + \alpha + \delta$, where δ may be positive or negative. The parameter δ is called the **dominance deviation**. When it is 0, there is no dominance, and the general model reduces to the additive model. In this notation,

TABLE 11.6 Breeding values for the additive model

Parental genotype	Breeding value (bv)
AA	$2f_a\alpha$
Aa	$-(f_A - f_a)\alpha$
aa	$-2f_A\alpha$

$$\bar{x} = x_{aa} + 2f_A\alpha + 2f_A f_a\delta \tag{11.16}$$

The breeding values are obtained by substituting into the expressions in Table 11.5 to obtain **Table 11.7**.

Although the expressions for the breeding values appear to be more complicated, we see in Table 11.7 that the results that are quite similar to those in Table 11.6. The only difference is that in Table 11.7, α is replaced by $\alpha' = \alpha - (f_A - f_a)\delta$. Therefore, by substituting in Equation 11.15, we obtain

TABLE 11.7 Breeding values for the general model of one QTL

Parental genotype	Breeding value (bv)
AA	$2f_a[\alpha-(f_A-f_a)\delta]$
Aa	$-(f_A-f_a)[\alpha-(f_A-f_a)\delta]$
aa	$-2f_a[\alpha-(f_A-f_a)\delta]$

$$V_A = 2f_A f_a \alpha'^2 = 2f_A f_a [\alpha-(f_A-f_a)\delta]^2 \quad (11.17)$$

for the general one-locus model of a QTL.

Unless $\delta = 0$, V_A is less than V_G. When there is some dominance, not all of the genetic variance is additive. After subtraction and a little more algebra, we find the difference to be

$$V_D = V_G - V_A = 4f_A^2 f_a^2 \delta^2 \quad (11.18)$$

We denote the difference by V_D because it represents the contribution of dominance to the total variance. If there is no dominance ($\delta = 0$), then the total genetic variance is the additive variance, as we have seen.

Multiple Quantitative Trait Loci

Quantitative characters are affected by more than one QTL. In the simplest situation, the overall genetic effect on a character is the sum of contributions from each. Even if that assumption is not correct, it is a good starting point. To illustrate, we will assume only two QTLs, A/a and B/b. As in the one-locus model, we pick one genotype, *aabb*, and describe the effects of other genotypes in terms of differences from the average of *aabb*. Adapting the notation from the previous section, we express the mean phenotype associated with each genotype in terms of α's and δ's for each locus, as in **Table 11.8**. Note that the mean of x for a genotype is not the sum of the means for each one-locus genotype. Instead, the differences from x_{aabb} are added. For example, the mean of x in individuals with genotype *AaBB* is

TABLE 11.8 Two-locus model in which effects of the two loci can be added

	AA	Aa	aa
BB	$x_{aabb}+2\alpha_A+2\alpha_B$	$x_{aabb}+\alpha_A+\delta_A+2\alpha_B$	$x_{aabb}+2\alpha_B$
Bb	$x_{aabb}+2\alpha_A+\alpha_B+\delta_B$	$x_{aabb}+\alpha_A+\delta_A+\alpha_B+\delta_B$	$x_{aabb}+\alpha_B+\delta_B$
bb	$x_{aabb}+2\alpha_A$	$x_{aabb}+\alpha_A+\delta_A$	x_{aabb}

x_{aabb} plus the deviation associated with the genotype at the first locus, $\alpha_A + \delta_A$, and the deviation associated with the genotype of the second locus, $2\alpha_B$. If the contributions of the two loci can be added, the mean and both the additive and dominance variances are the sum of the contributions of each locus, provided that the two loci are in linkage equilibrium. That is,

$$\bar{x} = x_{aabb} + 2f_A\alpha_A + 2f_Af_a\delta_A + 2f_B\alpha_B + 2f_Bf_b\delta_B \tag{11.19}$$

$$V_A = 2f_Af_a[\alpha_A - (f_A - f_a)\delta_A]^2 + 2f_Bf_b[\alpha_B - (f_B - f_b)\delta_B]^2 \tag{11.20}$$

and

$$V_D = 4f_A^2 f_a^2 \delta_A^2 + 4f_A^2 f_a^2 \delta_B^2 \tag{11.21}$$

When contributions from different loci cannot be added, then we say there is **epistasis** between the loci. Suppose, for example, that the average phenotype of an *AaBB* individual is $x_{aabb} + \alpha_A + \delta_A + 2\alpha_B + e_{AaBB}$. The last term represents the deviation from the sum of the effects of the two loci. The effect of epistasis on the genetic variance is similar to the effect of dominance. The average effect can be defined in terms of deviations from additivity (*e*) in the same way that α' is defined in terms of the dominance deviation δ. If there is epistasis among multiple QTLs, the algebra becomes quite daunting and we will not present it here. The principle is the same as with dominance. Epistasis affects the additive and dominance components of the genetic variance and creates a new component of the genetic variance, the epistatic variance. The total genetic variance is the sum of the additive, dominance, and epistatic variances. The interaction variance, V_I, in Equation 11.3 is the sum of the dominance and epistatic variances.

The term epistasis in this context can be confusing to people who have studied transmission genetics, where epistasis describes a specific kind of gene interaction in which the expression of one gene depends on the prior expression of another. To distinguish between the two uses of the term, *physiological epistasis* is sometimes used for epistasis in transmission genetics and *statistical epistasis* for epistasis in quantitative genetics.

Genotype–Environment Interactions

We have presented the basic elements of quantitative genetics as simply as possible. There are many complications that make application of the basic theory difficult. One is that the effect of a QTL may depend on environmental conditions. For example, human height, like many quantitative characters, is affected by nutrition during growth. The average height of people in Norway increased by 7.5 cm between 1855 and 1959, not because there was a significant change in the frequencies of alleles affecting height in the Norwegian population, but because the diet improved. If the average effects of all the genotypes of a QTL are changed in the same way because of a change in environment, then the additive genetic variance is remains the same. The only effect of the environmental change is to alter the reference average, x_{aa} (or x_{aabb}), which does not appear in the expressions for any of the variances. If, however, different genotypes respond differently to an environmental change, then the variances depend on the environment, and there is **genotype–environment interaction**, often abbreviated **G×E**. The theory of such interactions is complicated, but is based on the same principles we have already developed. We mention genotype–environment interactions to emphasize that any given quantitative genetic analysis is valid only for a particular population in a particular environment.

Mapping Quantitative Trait Loci

One of the goals of modern research in quantitative genetics is the mapping and identification of QTLs of importance to evolutionary biologists, plant and animal breeders, and medical geneticists. QTL mapping is relatively easy when controlled breeding can be done. The experimental design often used is a modification of Mendel's. The starting point is two populations that differ in the average of a quantitative character and that have been made homozygous for different alleles at a set of marker loci whose map positions are known. The two lines are crossed to form the F_1 generation and the F_1 is backcrossed to one of the parental populations. If a QTL that is partly responsible for the difference between the starting populations is closely linked to a marker locus, then in the backcross, different genotypes of the marker locus will have different mean values of the character. The theory is sketched in **Box 11.3**. This method has been used with *Drosophila melanogaster* because it is relatively easy to create populations homozygous at many marker loci. One study of QTLs that affect the number of abdominal bristles, illustrated in **Figure 11.5**, identified several genomic regions that affected average bristle number. One region on the X chromosome increased the average number by 2.81 in males and 5.12 in females. Abdominal and other bristles are often studied in *D. melanogaster* because they can be counted accurately, if somewhat tediously; in such a small organism, the measurement of continuously variable characters such as length is more error-prone.

In many species, including humans, controlled breeding is not possible. Genome-wide association studies (GWAS), described in Chapter 6, have been used extensively to map QTLs. A quantitative character is measured in a group of individuals who are also genotyped at a large number of SNPs. Then for each SNP, the averages of the individuals with the three genotypes are computed and tested for significant differences. The data shown in Table

Figure 11.5 *D. melanogaster* abdominal bristles.

BOX 11.3   Mapping Alleles When Starting with Homozygous Populations

We illustrate the basic idea of mapping using a single marker locus linked to a QTL. In practice, QTL mapping is done by analyzing data for numerous marker loci at the same time. In our case, the starting point is two populations that have been made homozygous by several generations of brother–sister mating. Assume that a QTL has two alleles, A and a, and that the average phenotypes associated with each genotype are x_{aa}, $x_{aa} + \alpha$, and $x_{aa} + 2\alpha$. That is, the additive effect of A is α, and there is no dominance. A linked marker locus with alleles B and b has no effect on the character. The recombination frequency between the two loci is c.

One of the two starting populations is homozygous for A and B and the other is homozygous for a and b. The F_1 is all $AaBb$, and the backcross to the $aabb$ population contains both parental and recombinant genotypes, as shown in the table.

Genotype	Frequency	Mean
$AaBb$	$(1-c)/2$	$x_{aa}+\alpha$
$Aabb$	$c/2$	$x_{aa}+\alpha$
$aaBb$	$c/2$	x_{aa}
$aabb$	$(1-c)/2$	x_{aa}

Only the genotype of the marker locus can be distinguished. The average in Bb individuals is $x_{aa} + \alpha(1-c)$, and the average in bb individuals is $x_{aa} + \alpha c$. If $c < ½$, then these averages will differ. Therefore, a test for linkage between a marker and a QTL is a test of a significant difference between the average phenotypes associated with different genotypes of a marker locus.

11.3 were obtained that way. For human height, more than 180 QTLs have been identified by this means.

Quantitative genetics will continue to be an active area of research. Identifying QTLs is the first step to understanding the how allelic differences at individual loci affect the phenotype. As the genomes of more species are sequenced, high quality genetic maps will become available for wider variety of organisms. The overall goal will be to connect our understanding of gene interactions at the molecular level to phenotypes of evolutionary and medical importance.

References

Becker W. A., 1992. *Manual of Quantitative Genetics*. Academic Enterprises, Pullman, WA.

Dudley J. W., 2007. From means to QTL: The Illinois long-term selection experiment as a case study in quantitative genetics. *Crop Science* 47: S20–S31.

*Falconer D. S. and Mackay T. F. C., 1996. *Introduction to Quantitative Genetics*. Longman, Essex.

Galton F., 1889. *Natural Inheritance*. MacMillan & Sons, London.

*Lynch M. and Walsh B., 1998. *Genetics and Analysis of Quantitative Traits*. Sinauer Associates, Sunderland, MA.

Long A. D., Mullaney S. L., Reid L. A., et al., 1995. High resolution mapping of genetic factors affecting abdominal bristle number in *Drosophila melanogaster*. *Genetics* 139: 1273–91.

Weber K. E. and Diggins L. T., 1990. Increased selection response in larger populations. 2. Selection for ethanol vapor resistance in *Drosophila melanogaster* at two population sizes. *Genetics* 125: 585–597.

Weedon M. N., Lettre G., Freathy R. M., et al., 2007. A common variant of *HMGA2* is associated with adult and childhood height in the general population. *Nature Genetics* 39: 1245–1250.

*Recommended reading

EXERCISES

11.1 Suppose you had a sample of 10 adult individuals from a population of mice and found their weights to be 81, 82, 84, 85, 87, 91, 91, 93, 94, and 95 g. What are the mean, variance, and standard deviation of their weights?

11.2 Suppose that the weights of the five male parents in Table 11.2 are (A) 650, (B) 670, (C) 700, (D) 630, (E) 680.

a. Use the data in the first line of Table 11.2 to estimate the additive variance by computing the parent–offspring covariance.

b. Then, assume each of the five males was mated to only one female and the offspring weights are given by the first line (687, 618, etc.). What is the heritability? (Use the variance of the fathers' weights to estimate the population variance.)

11.3 Suppose that a population of chickens has 6-week weights of 691, 680, 687, 657, 658, 793, 592, 763, 669, and 374 g.

a. With the heritability you computed in Exercise 11.2, use the breeder's equation to predict the average size of the offspring if the five largest chickens are chosen as parents.

       b. Predict the average size if the two largest chickens are chosen as parents.

       c. Although your results from parts a and b indicate that choosing a smaller number of the largest parents will always produce the more rapid response to selection, it is not usually good to select only a few parents from each generation. Why?

11.4 For the data in Table 11.3, compute α and δ for males and for females.

11.5 Confirm that the average breeding value in the additive model is 0.

11.6 For the model illustrated in Figure 11.5. compute $\bar{x}$, V_x and V_A.

11.7 Suppose a QTL is recessive in its effect on a quantitative character, thus $x_{Aa} = x_{aa}$. Express V_A and V_D in terms of α and f_A and plot V_A and V_D as functions of f_A.

11.8 It is possible that, for a QTL, the average heterozygote is larger than the larger homozygote. That is, $\delta > \alpha$. In this case, find the frequency of A at which $V_A = 0$ but $V_D > 0$.

11.9 Suppose that a quantitative character is governed by two loci with two alleles each, A/a and B/b, and suppose that all of the genotypes have the same average phenotype, x_{aabb} except for $AABB$ individuals, which have an average phenotype $x_{aabb} + \alpha$. Find the average of the character and the total genetic variance of the character if the two loci are in Hardy–Weinberg and linkage equilibrium with allele frequencies $f_A = 0.7$ and $f_B = 0.8$, $x_{aabb} = 10$ and $\alpha = 1$.

Appendix A

Basic Probability Theory

In Box 1.1 of the main text we introduce the concept of probability. Here, we will expand upon those ideas. While the book can be read without knowledge of the material covered in this appendix, students who have read this appendix may find many topics covered throughout the book easier to digest.

The first concept we will introduce is the idea of **conditional probability**. Conditional probability is used to express the belief in future events conditional on some information. For example, consider an experiment with tosses of a fair die, with sample space {1, 2, 3, 4, 5, 6} and associated probabilities:

$$\Pr(i) = 1/6, i = 1, 2, \ldots, 6$$

A student casts the die, but we are only told that the result of the cast was less than 4. What is the probability that the result was 1? We can express this probability as

$$\Pr(die\ shows\ 1 \mid die\ shows\ number < 4)$$

Notice here that the vertical bar means "given that." We read the above statement as "the probability that the die shows 1 given that the die shows a number less than 4." We might intuitively guess that the answer to this question is 1/3, since there are three possible outcomes less than 4, each of which may occur with equal probability. This answer, and answers to other questions involving probabilities, can be derived rigorously from the three basic **laws (axioms) of probability**, which are:

1. For any event E,

$$0 \le \Pr(E) \le 1 \text{ and } \Pr(E \mid E) = 1$$

2. If two events E and A are <u>mutually exclusive</u>, then

$$\Pr(E \text{ or } A) = \Pr(E) + \Pr(A)$$

Mutually exclusive events are ones for which we know that if one happens, then none of the others can happen. For example, for the coin toss example mentioned in Box 1.1, $\Pr(H \text{ and } T) = 0$, i.e., the two events are mutually exclusive. A coin cannot land on its head and its tail simultaneously in the same toss. Often the notation $\Pr(H, T)$ is used instead of $\Pr(H \text{ and } T)$ to denote the probability of both H and T occurring.

3. For two events, E and A, with $\Pr(A) > 0$,

$$\Pr(E \mid A) = \Pr(E, A) / \Pr(A)$$

implying that

$$\Pr(E, A) = \Pr(E \mid A)\Pr(A)$$

The concept of <u>**independence**</u> is of great importance in probability theory. Two events, E and A, are independent if

$$\Pr(E, A) = \Pr_1(E)\Pr_1(A)$$

From the third law of probability, we see that independence of E and H implies that $\Pr(E \mid A) = \Pr(E)$ and $\Pr(A \mid E) = \Pr(A)$.

The **law of total probability** states that if A_i, $i = 1, 2, \ldots, r$, are mutually exclusive events and $\sum_{i=1}^{r} \Pr(A_i) = 1$, then for any event E

$$\Pr(E) = \sum_{i=1}^{r} \Pr(E \mid A_i)\Pr(A_i)$$

Under the same conditions (and assuming $\Pr(E) > 0$), Bayes' Theorem states that

$$\Pr(A_j \mid E) = \frac{\Pr(E, A_j)}{\Pr(E)} = \frac{\Pr(E \mid A_j)\Pr(A_j)}{\Pr(E)} = \frac{\Pr(E \mid A_j)\Pr(A_j)}{\sum_{i=1}^{r} \Pr(E \mid A_i)\Pr(A_i)}$$

The Binomial RV

In Box 1.1 we introduced the concept of a random variable (RV). We used a coin-toss random variable with sample space $\{H, T\}$ as an example. Much can be learned about basic probability theory by gaining familiarity with some standard RVs and their properties. RVs come in two "flavors," discrete and continuous. We will start by discussing some discrete RVs. <u>Discrete RVs</u> are defined by a **probability mass function (PMF)**. The PMF assigns probability to each possible event (elements of the state space). The first RV we will discuss is the **Bernoulli RV**, which is defined by the following PMF:

PMF: Bernoulli

$$\Pr(X = x) = \begin{cases} p & \text{if } x = 1 \\ 1-p & \text{if } x = 0 \\ 0 & \text{otherwise} \end{cases}, \qquad 0 \le p \le 1$$

The previously discussed fair coin toss is an example of a Bernoulli RV with $p = \frac{1}{2}$, if we code H as 1 and T as 0.

The next RV we will be interested in is the **binomial RV**. The binomial RV has sample space $\{0, 1, \ldots, n\}$ and PMF:

PMF: Binomial

$$\Pr(X = x) = \binom{n}{x} p^x (1-p)^{n-x}, \qquad 0 \le p \le 1$$

The notation $\binom{n}{x} = n!/(x!(n-x)!)$ is read "n choose x" and can be thought of as the number of different ways you can sample x balls from a bag with a total of n balls. The notation $n!$ is the factorial function, defined as $n \times (n - 1) \times (n - 2) \times \ldots \times 1$ (**Figure A.1**).

The binomial random variable describes the sum of n independent Bernoulli RVs. To show this, let us first consider the case of two independent Bernoulli RVs X_1 and X_2 with common parameter p. Then, by the definition of independence and the second law of probability,

$$\Pr(X_1 + X_2 = 1) = p(1 - p) + (1 - p)p = 2(1 - p)p$$

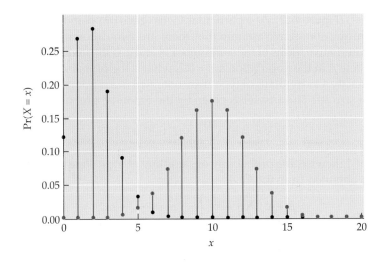

Figure A.1 The binomial distribution for $n = 20$ and $p = 0.5$ (red dots) and $p = 0.1$ (black dots).

In other words, the probability of observing exactly one head in two coin tosses is $2(1-p)p$, where p is the probability of observing heads in a toss.

Notice that since $\binom{2}{1} = 2$, this matches the definition of the binomial RV.

Now consider n independent Bernoulli RVs with common parameter p corresponding to n tosses of a (possibly biased) coin. We are interested in the sum of these RVs. We notice that if the sum of these RVs adds to j, this is identical to tossing a coin n times and observing heads exactly j times and tails exactly $n-j$ times. This could occur, for example, by first obtaining j heads and then $n-j$ tails with probability $p^j(1-p)^{n-j}$. Of course, there are many other possible ways to get the desired outcome. For example, we could observe a tails first, then j heads and then $n-j-1$ tails, etc. The total number of ways we could observe j heads is $\binom{n}{j}$, and each of these outcomes occurs with probability $p^j(1-p)^{n-j}$.

Therefore,

$$\Pr(X_1 + X_2 + \ldots + X_n = j) = \binom{n}{j} p^j (1-p)^{n-j}$$

which is identical to the definition of the binomial RV.

The **cumulative distribution function (CDF)** of a random variable (X) is defined as $F_X(x) = \Pr(X \leq x)$.

For the binomial RV, it equals

$$F_X(x) = \sum_{i=0}^{x} \binom{n}{i} p^i (1-p)^{n-i}, \qquad 0 \leq p \leq 1$$

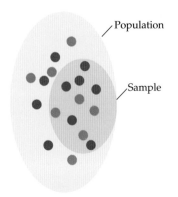

Figure A.2 A sample of 10 individuals with 4 copies of the A allele (blue) and 6 copies of the a allele (red).

EXAMPLE 1 Consider a large haploid population (one in which each individual only has one copy of the genetic material) (**Figure A.2**). At a particular locus (gene or position in the chromosome), we assume that there are two different alleles (alternative forms of the same gene): A and a. If the frequency of allele A in the population is $p = 0.2$, what is the probability that exactly 4 individuals of type A are obtained in a random sample of size 10? The answer is given by the binomial probability

$$\Pr(X = 4) = \binom{10}{4} 0.2^4 0.8^6 \approx 0.088$$

Likewise, the probability of sampling at most 4 individuals is obtained from the CDF:

$$F_X(4) = \sum_{i=0}^{4} \binom{10}{i} 0.2^i 0.8^{10-i} \approx 0.967$$

EXAMPLE 2   In the following we will consider a population genetic model. We will assume an idealized population of $2N$ randomly mating haploid individuals, that is, individuals with exactly one copy of genetic material, that evolve in discrete generations. The Wright–Fisher model asserts that the number of offspring of any individual in the next generation is binomially distributed, with parameters $n = 2N$, and $p = 1/(2N)$. This is a natural assumption if all individuals have the same fitness (potential for reproductive success). It is identical to assuming that all individuals in generation t have equal probability of being the parent of an individual in generation $t + 1$ and that this probability is independent among all off-spring. If the probability that any particular offspring in generation $t + 1$ is a descendent of

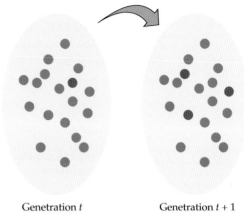

Figure A.3 Wright–Fisher sampling between two generations. One individual (red) has three offspring in the next generation.

Genetration t Genetration $t + 1$

individual i in generation t is governed by a Bernoulli RV with parameter $p = 1/(2N)$, then assuming independence, the total number of descendants of individual i in the next generation is a binomial RV. This follows from fact that the sum of independent Bernoulli RVs is binomially distributed.

In the example shown in **Figure A.3**, the individual in generation t labeled as a red circle had three offspring in the next generation (the three red circles). The probability of this event is $\begin{pmatrix} 2N \\ 3 \end{pmatrix}[1/(2N)]^3[1-1/(2N)]^{2N-3}$.

If the population size is $2N = 20$ (as indicated by the dots in Figure A.2), this equals approximately 0.0596.

Expectation

The concept of expectation was introduced in Box 2.1 of the main text. The expectation of a discrete RV is defined as

$$E[X] = \sum_x x \Pr(X = x)$$

where the sum is over all elements of the sample space. Notice that the expectation corresponds to an average value of the RV. The expectation for the Bernoulli RV is

$$E[X] = 1 \times p + 0 \times (1 - p) = p$$

One of the most important properties of expectations is that the *expectation of a sum* of RVs is equal to the sum of the expectations, that is,

$$E[X + Y] = E[X] + E[Y]$$

We can derive the expectation of the binomial RV from this rule using the fact that the binomial RV is a sum of independent Bernoulli RVs. Let Y be distributed binomially with parameters n and p and let X_i, $i = 1, 2, \ldots, n$, be independent Bernoulli RVs with parameter p. Then

$$Y = X_1 + X_2 + \ldots + X_n$$

and

$$E[Y] = E[X_1 + X_2 + \ldots + X_n] = E[X_1] + E[X_2] + \ldots + E[X_n] = nE[X_i] = np$$

The concept of expectation can be extended to the *expectation of a function of an RV* as follows:

$$E[f(X)] = \sum_x f(x) \Pr(X = x)$$

where $f(X)$ is any function of X. For a linear function, $f(X) = aX + b$, it follows that

$$E[aX + b] = \sum_x (ax + b) \Pr(X = x) = a \sum_x x \Pr(X = x) + b \sum_x \Pr(X = x) = aE[X] + b$$

Since any conditional probability is a real probability in the sense that it obeys the basic axioms of probability, we can define a *conditional expectation*:

$$E[X \mid Y = y] = \sum_x x \Pr(X = x \mid Y = y)$$

EXAMPLE 3   If individuals with certain alleles leave more offspring in the next generation than individuals without them, this is known as selection. Assume that there are two segregating alleles in the population, A and a. The **fitness** of the two alleles can be expressed as w_A and w_a. Fitness is defined such that the expected frequency of allele A in the next generation is

$$E[p_{t+1} \mid p_t] = \frac{p_t w_A}{p_t w_A + (1 - p_t) w_a}$$

The distribution of the number of A alleles in the next generation is binomially distributed with parameters $E[p_{t+1} \mid p_t]$ and $2N$. What is the probability that a new allele (of type A in a background of type a alleles) arising in the population, which increases the fitness by 10%, will get lost after the very first generation? We see that

$$E[p_{t+1} \mid p_t] = \frac{[1/(2N)] \times 1.1}{[1/(2N)] \times 1.1 + [1 - 1/(2N)]}$$

So the probability of immediate loss is

$$\left(\frac{[1/(2N)] \times 1.1}{[1/(2N)] \times 1.1 + [1 - 1/(2N)]} \right)^{2N}$$

For large population sizes, this equals about 0.333. So we see that even though the allele provided a strong selective advantage (increase in fitness), it has a high probability of being immediately lost. Most beneficial mutations in nature probably confer much lower fitness, implying that a large proportion of them will never survive the initial genetic drift.

We have here modeled selection for a haploid population. The results using such a model are identical to the results obtained for a diploid population with genic selection (see Box 7.5).

Variance

The **variance** of a random variable (X) is defined as

$$V[X] = E[(X - E[X])^2]$$

Notice that the variance is just a type of expectation (or expectation of a function of a RV)—the expectation of the squared deviation from the mean (expectation) of a random variable. Several important properties of variance can easily be derived from its definition:

$$V[X] = E[(X - E[X])^2] = E[X^2 - 2XE[X] + E[X]^2]$$
$$= E[X^2] - 2E[X]E[X] + E[X]^2 = E[X^2] - E[X]^2$$

We can use this expression to derive the variance of the Bernoulli RV. For the Bernoulli RV, we know that $E[X] = p$. We also have

$$E[X^2] = 1^2 \times p + 0^2 \times (1 - p) = p$$

So the variance of the Bernoulli RV is

$$V[X] = p - p^2 = p(1 - p)$$

Another useful property of variances is the simple formula for the variance of a linear function of a RV. From the definition of variance, we have

$$V[a + bX] = E[\{(a + bX) - E[a + bX]\}^2] = E[\{a + bX - a + bE[X]\}^2]$$
$$= E[\{b(X - E[X])\}^2] = b^2E[(X - E[X])^2] = b^2V[X]$$

Finally, for independent RVs, the variance of the sum is just the sum of the variances:

$$V[X + Y] = V[X] + V[Y]$$

It is important to remember that this is true if and only if X and Y are independent. This result can be used to derive the variance for a binomial RV, remembering that a binomial RV is a sum of independent Bernoulli RVs. Let Y be a distributed binomial with parameters n and p, and let X_i, $i = 1,2,\ldots,n$ be independent Bernoulli RVs with parameter p. Then

$$V[Y] = V[X_1 + X_2 + \ldots + X_n] = nV[X_i] = np(1 - p)$$

EXAMPLE 4 Knowing the variance of a binomial RV, we can now easily derive the variance of the allele frequency in the Wright–Fisher model due to genetic drift. Using the notation and definitions from Example 3, we have $E[p_{t+1} \mid p_t] = p_t$ and $p_t = n_t/(2N)$. Then

$$V[p_{t+1} \mid p_t] = V[n_{t+1}/2N \mid p_t] = \frac{1}{2N^2}V[n_{t+1} \mid p_t] = \frac{1}{2N^2}2N(1-p_t)p_t = \frac{(1-p_t)p_t}{2N}$$

The magnitude of the variance governs the strength of genetic drift (how fast allele frequencies change through time). We see that as the population size increases, the variance decreases—so genetic drift is most efficient (fast) in small populations.

The Poisson RV

The **Poisson RV** has sample space on {0, 1, 2, . . .} and is defined by the following PMF:

PMF: Poisson

$$\Pr(X = x) = \lambda^x e^{-\lambda}/x!, \qquad 0 \le \lambda$$

It is often used to model the number of events occurring in a time interval. The Poisson distribution can be used to approximate the binomial distribution. Consider a binomial RV (X) with PMF

$$\Pr(X = x) = \binom{n}{x} p^x (1-p)^{n-x}, \qquad 0 \le p \le 1$$

Now considering the limit of $n \to \infty$, $p \to 0$ while $np \to \lambda$, from calculus we can show

$$\Pr(X = x) \to \lambda^x e^{-\lambda}/x!$$

which is exactly the PMF for the Poisson distribution. So the Poisson distribution approximates the binomial distribution if many Bernoulli trials occur (n is large) but each has a very small success probability (p is small).

EXAMPLE 5 Consider a Wright–Fisher population as in Example 2. Remember that the number of offspring of an individual is binomially distributed with parameters $2N$ and $1/(2N)$. If the population is very large, so that $2N \to \infty$, then $1/(2N) \to 0$, but $2N \times 1/(2N) \to 1$, so the number of offspring of an individual is approximately Poisson-distributed with parameter 1. What is the probability that an individual has exactly 3 offspring for a population of size 20? Using the Poisson approximation we find

$$1^3 e^{-1}/3! \approx 0.0613$$

The exact binomial probability was found in Example 2 to be 0.0596. So for $2N = 20$, the Poisson approximation is (in this case) not very precise. However, for more realistic population sizes, it will be very precise. For a population of size 200, the exact binomial sampling probability is approximately 0.0612.

EXAMPLE 6 Consider again the Wright–Fisher model. We will add mutation to the model by assuming that each gene copy mutates with probability μ in each generation. A mutation causes one allele to change into another allele. We are now interested in tracking the history of a gene copy through

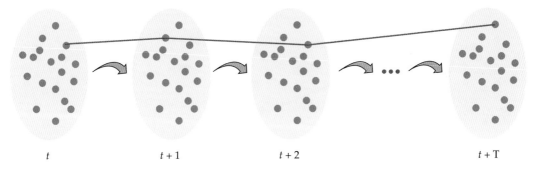

Figure A.4 Wright–Fisher sampling between multiple generations. The ancestry of one individual is shown as a red line.

generations (**Figure A.4**). In particular, we want to answer the following question: during the last T generations, what is the probability that k mutations occurred in the ancestry of the gene copy? We see that the chance of observing k mutations is given by a binomial RV:

$$\Pr(\textit{Number of mutations} = k) = \binom{T}{k} \mu^k (1-\mu)^{T-k} \approx (\mu T)^k e^{-\mu T} / k!$$

assuming that T is large and μ is small. So the number of mutations in a lineage of length T (the ancestry of a single gene copy during T generations) is approximately Poisson distributed with parameter μT. This fact is used extensively in coalescence theory.

The expectation of a Poisson RV is easily derived from the definition of expectation. Let X be a Poisson RV with parameter λ, then

$$E[X] = \sum_{k=0}^{\infty} k \lambda^k e^{-\lambda} / k! = \lambda e^{-\lambda} \sum_{k=1}^{\infty} \lambda^{k-1} / (k-1)! = \lambda e^{-\lambda} \sum_{j=0}^{\infty} \lambda^j / j! = \lambda e^{-\lambda} e^{\lambda} = \lambda$$

The variance of the Poisson RV is also $V[X] = \lambda$.

The Geometric RV

The **geometric RV** has sample space on $\{1, 2, \ldots\}$ and PMF:

PMF: Geometric

$$\Pr(X = x) = p(1 - p)^{x-1}, \qquad 0 < p < 1$$

We can think of the geometric RV as describing the distribution of the time (in number of trials) to the first success in a series of independent Bernoulli trials each with success probability p. Keep tossing a coin until heads appears for the first time. If the probability of heads is p, then the distribution of the number of coin tosses follows a geometric RV.

The expectation of the geometric RV is $1/p$ and the variance is $1/p^2$.

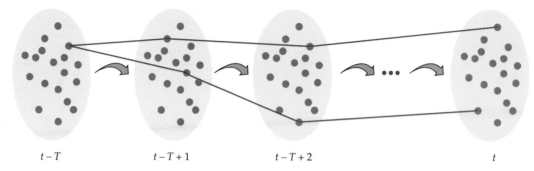

$t - T$ $t - T + 1$ $t - T + 2$ t

Figure A.5 Wright–Fisher sampling between multiple generations tracking the ancestry of two gene copies (haploid individuals) back in time until they find a most recent common ancestor.

EXAMPLE 7 In a sample of two gene copies taken from a population, what is the distribution and expectation of the time until the two gene copies find a most recent common ancestor in their ancestry (T)? T is also called the *coalescence time* (**Figure A.5**).

Let the current generation be generation t. For the two gene copies sampled at generation t to have a most recent common ancestor in generation $t - 1$, they must both be descendants of the same parental gene copy in generation $t - 1$. Because all gene copies in generation $t - 1$ have the same probability of being parental to a gene copy in generation t, this happens with probability $1/(2N)$. So $\Pr(T = 1) = 1/(2N)$.

For two gene copies to have the most recent common ancestor in generation $t - 2$, they must first not have a most recent ancestor in generation $t - 1$, and that happens with probability $1 - 1/(2N)$. They must then be descendants of the same parental gene copy in generation $t - 2$, and that happens with probability $1/(2N)$. So $\Pr(T = 2) = [1 - 1/(2N)] \times 1/(2N)$. Continuing this logic for larger values of T, we see that

$$\Pr(T = x) = [1 - 1/(2N)]^{x-1} 1/(2N)$$

We recognize this as a geometric random variable with parameter $1/(2N)$. The expected coalescence time in a sample of size 2 is $2N$ (the population size).

EXAMPLE 8 Assume you sample two genes copies from a Wright–Fisher population. What is the probability that no mutations occurred in the history of two sampled gene copies before their most recent common ancestor? This is an important question in population genetics. If no mutations happened, this implies that the two genes copies are **identical by descent (IBD)**. Assuming that there are no back-mutations (e.g., mutations from allele A to allele a and back again to allele A), this is also the probability of two gene copies being identical—the *expected homozygosity*.

Two gene copies are identical by descent if no mutations happened in the previous generation [with probability $(1 - \mu)^2$] and if the two gene copies either found a most recent common ancestor in the previous generation [with probability $1/(2N)$] or if the two different ancestral gene copies in the previous generation were IBD, i.e.,

$$\Pr(IBD) = (1 - \mu)^2\{[1/(2N)] + [1 - 1/(2N)]\Pr(IBD)\}$$

Rearranging, we find:

$$\Pr(IBD) = \frac{(1-\mu)^2}{2N\{1-[1-1/(2N)](1-\mu)^2\}} = \frac{1-2\mu+\mu^2}{1+2N\mu-2\mu-N\mu^2+\mu^2}$$

As often in population genetics, we now consider the limit of large population sizes and small mutations rates, i.e., we let $2N \to \infty$, $\mu \to 0$ so that $2N\mu \to \theta$. Then

$$\Pr(IBD) \to \frac{1}{1+\theta}$$

The probability of identity by descent is approximately $1/(1 + \theta)$. The same result is derived in the text using the exponential distribution and an infinite alleles model.

The CDF of the geometric RV is given by

$$F_X(x) = \sum_{i=1}^{x} p(1-p)^{i-1} = 1-(1-p)^x$$

Appendix B

The Exponential Distribution and Coalescence Times

The exponential random variable is a **continuous random variable**. In contrast to, say, the binomial random variable (Appendix A), the outcomes are not integers, but can take any non-negative real number; thus, the sample space of the exponential random variable is $\mathbb{R}^+$ (all positive real numbers). Continuous random variables are defined in terms of **probability density functions** (**PDFs**). For the exponential random variable, the PDF is

$$f(x) = \lambda e^{-\lambda x}, \quad 0 < \lambda$$

The parameter λ (lambda) is the rate. The mean of the exponential is $1/\lambda$. In the main text, we use the example of waiting for an empty taxi in Manhattan. The exponential random variable can be used to model the time we have to wait for the first taxi, if taxis arrive independently of each other at a constant rate (λ). The function $f(x)$ is not a probability, but a density. If $f(a)$ is three times as large as $f(b)$, it means that the chance that the first taxi will arrive *around* time a is three times as large as the chance that it will arrive *around* time b. But $f(a)$ is *not* the probability that the first taxi arrives *at* time a. Because we measure time continuously, there are infinitely many possible values, and the chance that any particular value is the true one (when measured with absolute precision) is zero. However, we can obtain probabilities for time intervals by integrating over the PDF. The integral $\int_a^b f(x)\,dx$ gives the probability hat the taxi arrives in the time interval between time a and b.

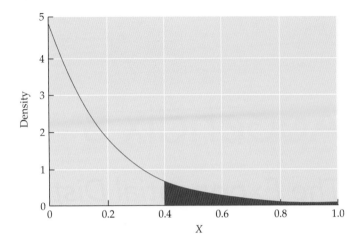

Figure B.1 The PDF of a random variable with parameter $\lambda = 5$. The shaded area gives the probability that the random variable takes on a value > 0.4.

A plot of the PDF on an exponential distribution with parameter $\lambda = 5$ is shown in **Figure B.1**.

What is $\Pr(X > 0.4)$ for this example? We realize that this probability equals the surface of the shaded area in Figure B.1, that is,

$$\int_{0.4}^{\infty} 5e^{-5x}\,dx = -e^{-5x}\Big|_{0.4}^{\infty} = 0 - \left(-e^{-5\times0.4}\right) \approx 0.135$$

In the main text, we show that the probability of no coalescence events (no common ancestor) before time t in a sample of size $n = 2$ is e^{-t}. The probability that an exponential random variable with rate $\lambda = 1$ takes on a value less than t is $\int_0^t e^{-x}\,dx = 1 - e^{-t}$. So the probability that the value is larger than t is $1 - (1 - e^{-t}) = e^{-t}$. The identity between these two expressions shows that the distribution of the time to the first coalescence event in a sample of size $n = 2$ is exponentially distributed with parameter $\lambda = 1$. We later argue that the distribution to the first coalescence time in a sample of n chromosomes also is exponentially distributed, but with rate $\lambda = n(n-1)/2$. Examples of densities of coalescence times is given in **Figure B.2**.

As an example, assume we know that a bottleneck (strong temporary decrease in population size) happened in a population 100 generations ago. We also believe that the population after that essentially has evolved as a Wright–Fisher population with $2N=500$. What is the probability that two gene copies sampled from the population coalesced before the time of the bottleneck? Using the continuous approximation (exponential distribution), we obtain the answer to this question as

$$\Pr(t \le 100/500) = \int_0^{1/5} e^{-x}\,dx = -e^{-x}\Big|_0^{1/5} = 1 - e^{-1/5} \approx 0.181$$

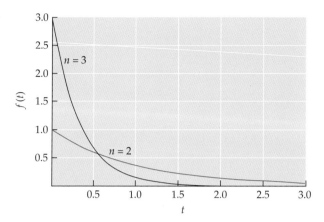

Figure B.2 The distribution of coalescence times when there are $n = 3$ lineages (an exponential distribution with mean 1/3) and $n = 2$ lineages (an exponential distribution with mean 1). Times is measured in terms of $2N$ generations. The area under the curve, calculated up to any particular value, gives the probability that the coalescence time is less than that value.

Two facts about exponential random variables are very useful in coalescence theory, and we will state them here without proof. First, if there are k independent exponentially distributed random variables, each representing a different type of event, and the rate of the ith variable is λ_i, $i = 1, 2, \ldots k$, then the waiting time to the first event (the minimum of the k random variables) is exponentially distributed with rate $\lambda_T = \sum_{i=1}^{k} \lambda_i$, that is, the mean waiting time is $1/\lambda_T$. Second, the probability that the first event that happens is an event of type j is λ_j/λ_T, i.e., it is the rate at which this type of event occurs divided by the total rate. Think again in terms of waiting for a taxi in Manhattan. If there are k taxi companies, the probability that the first empty taxi to arrive is from any particular taxi company is the rate at which taxis arrive from this company divided by the total rate at which taxis arrive. This might make intuitive sense to some readers.

Appendix C

Maximum Likelihood and Bayesian Estimation

Statistical estimation usually proceeds by first defining a set of assumptions (a model) and then some principle based on this set of assumptions is used to devise an estimator. For example, the Wright–Fisher model and the infinite sites model together provide a set of assumptions under which θ can be estimated. We saw that based on the statistic π, we could estimate θ. The statistical principle that is most commonly used to devise estimators is the principle of maximum likelihood (Chapter 5). The statistic used in maximum likelihood estimation is the **likelihood function**. The likelihood function is defined as any function proportional to the sampling probability of the data; thus, if X is discrete data and θ is the parameter, then $\Pr(X \mid \theta)$ is the likelihood function.

As an example, assume an infinite sites model and assume that two sequences have been sampled and that there were three nucleotide differences between the two sequences. What is the likelihood function for θ? From the chapter on coalescence theory, we know it is given by

$$L(\theta) = \frac{1}{1+\theta}\left(\frac{\theta}{1+\theta}\right)^3$$

Any other function that differs from this function only by a constant of proportionality is also a likelihood function. You can multiply the likelihood function by any positive constant, and it remains a valid likelihood function.

The maximum likelihood principle tells us that we should use the value of θ that maximizes the likelihood function $L(\theta)$ with respect to the parameter (θ). When the parameter is a simple scalar variable (such as θ), this is usually done using the following steps:

(1) Define the likelihood function, $L(\theta)$.

(2) Take the logarithm of the likelihood function, $\ell(\theta)$.

(3) Take the derivative of the likelihood function with respect to the parameter, $\ell'(\theta)$.

(4) Equate the derivative to zero [$\ell'(\theta) = 0$] and solve for the parameter to find $\hat{\theta}$.

(5) Confirm that $\hat{\theta}$ is in fact a maximum by checking that the second derivative of $\ell(\theta)$ evaluated at $\hat{\theta}$ is negative. Verify that the global maximum has been found.

EXAMPLE 1 For the example of three nucleotides in a comparison of two sequences, we find

(1) $L(\theta) = \dfrac{1}{1+\theta}\left(\dfrac{\theta}{1+\theta}\right)^3$ **(Figure C.1A)**

(2) $\ell(\theta) = 3Log(\theta) - 4Log(1+\theta)$ **(Figure C.1B)**

(3) $\ell'(\theta) = \dfrac{3}{\theta} - \dfrac{4}{1+\theta}$

(4) $\dfrac{3}{\hat{\theta}} - \dfrac{4}{1+\hat{\theta}} = 0 \Rightarrow \hat{\theta} = 3$

(5) $\ell''(\theta) = -\dfrac{3}{\theta^2} + \dfrac{4}{(1+\theta)^2}$

Because $\theta = 3$ is the only solution to the Equation defined in step 3, and because the second derivative is negative when $\theta = 3$, the log likelihood function, and therefore also the likelihood function, is maximized at $\theta = 3$.

Notice that the maximum likelihood estimate in this case corresponds to the expected number of substitutions and is identical to both Watterson's and Tajima's estimators of θ. In general, the maximum likelihood estimate will be identical to Watterson's and Tajima's estimate only for sample sizes of two sequences, not for larger samples.

EXAMPLE 2 We will discuss one more example. Assume a sample of size n is obtained from a population. The sample contains x copies of allele A and $n - x$ copies of allele a. What is the maximum likelihood estimate of the frequency of allele A in the population (p)? Assuming random sampling,

(A)

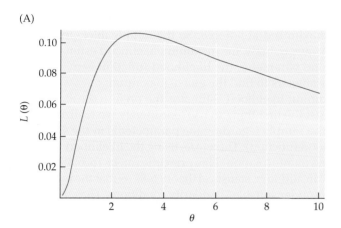

Figure C.1 The likelihood function (A) and log likelihood function (B) for θ under the standard coalescent with infinite sites mutation for two sequences with three nucleotide differences.

(B)

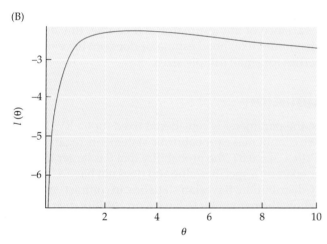

we identify the binomial distribution as the correct sampling distribution (likelihood function). We then have

(1) $L(p) = \begin{pmatrix} n \\ x \end{pmatrix} p^x (1-p)^{n-x}$

(2) $\ell(p) = x Log(p) + (n-x) Log(1-p)$

(3) $\ell'(p) = \dfrac{x}{p} - \dfrac{n-x}{1-p}$

(4) $\dfrac{x}{\hat{p}} - \dfrac{n-x}{1-\hat{p}} = 0 \Rightarrow \hat{p} = \dfrac{x}{n}$

(5) $\ell''(p) = -\dfrac{x}{p^2} - \dfrac{(n-x)}{(1-p)^2}$

Because the second derivative is negative when $p = x/n$, the likelihood function is in fact maximized at $p = x/n$. So the sample allele frequency (x/n) is, perhaps not surprisingly, the maximum likelihood estimator of p.

For more than one parameter, the likelihood function is maximized jointly for all the parameters to find the global maximum.

Why is the maximum likelihood estimator the preferred estimator in many cases (or at least the first thing to try)? Intuitively we realize that all the information in the sample regarding the parameter is in the sampling distribution (likelihood function). Therefore, using the likelihood function does not lead to a loss of information. This principle can be formalized and is called **sufficiency**. In addition, statistical theory tells us that the maximum likelihood estimator (under suitable "regularity conditions") has some very desirable properties such as **consistency** (the estimate converges to the true value of the parameter as the sample size gets large) and **asymptotic efficiency** (for large samples the maximum likelihood estimate has the minimum possible variance of any estimator). Of course, this does not guarantee that the maximum likelihood estimator does not have undesirable statistical properties for small sample sizes.

Bayesian Estimation

In classical statistics, the parameter is considered fixed (not a RV) but unknown. Therefore, we cannot make probabilistic statements regarding the parameter such as $\Pr(\theta > 3) = 0.5$ ("the probability that θ is larger than 3 is 0.5"). The parameter takes on a specific value and does not have a (non-degenerate) probability distribution. **Bayesian statistics**, in contrast, are based on the idea that we allow the parameter to have a probability distribution. We quantify our knowledge about the parameter before we have seen any data or conducted any experiments, in terms of a prior distribution, $\Pr(\theta = q)$ for a discrete RV, or $f_\theta(q)$ for a continuous RV. After having observed the data, or performed an experiment, our updated belief in the values that θ might take is quantified by a posterior distribution $\Pr(\theta = q \mid X = x)$ or $f_\theta(q \mid X = x)$. Using Bayes' Theorem, the posterior distribution is calculated as

$$f_\theta(q \mid X = x) = \frac{\Pr(X = x \mid \theta = q) f_\theta(q)}{\Pr(X = x)} = \frac{\Pr(X = x \mid \theta = q) f_\theta(q)}{\int \Pr(X = x \mid \theta = q) f_\theta(q) dq}$$

assuming that the data is discrete but the prior distribution for the parameter is continuous. Notice that the posterior distribution is given by the product of the likelihood function and the prior distribution multiplied by a scaling factor that does not depend on the value of the parameter, but does depend on the data. We can now make direct probabilistic statements about the parameter. For example, we can find the probability that the parameter is located in a certain interval. Such intervals are called **credible intervals** and take the same role as confidence intervals in classical statistics.

EXAMPLE 3 Under a standard coalescence model (see Chapter 3) and assuming an infinite sites model, what is the posterior distribution of the time to the most recent common ancestor (coalescence time) of two sequences, given that the two sequences are identical ($S = 0$)? The prior distribution of the coalescence time is $f(t) = e^{-t}$, the likelihood function is a Poisson, $\Pr(S = 0 \mid T = t) = e^{-\theta t}$, and the scaling factor is $\Pr(S = 0) = \left(\dfrac{1}{1+\theta}\right)$, the probability of observing no mutations. Now, from Bayes' Theorem,

$$f(t \mid S = 0) = \frac{f(t)\Pr(S = 0 \mid T = t)}{\Pr(S = 0)} = e^{-t}e^{-\theta t} \bigg/ \left(\frac{1}{1+\theta}\right) = (1+\theta)e^{-t(1+\theta)}$$

which is just an exponential distribution with parameter $1 + \theta$. Notice that in this case, the Bayesian framework is the natural framework for estimating the parameter because we already have a well-defined prior distribution.

EXAMPLE 4 Under the assumptions of Example 2, what is the posterior probability that the coalescence time is less than 1? Knowing that the posterior distribution is exponential with parameter $1 + \theta$, we can get the result directly from the CDF of the exponential random variable:

$$\Pr(T < 1 \mid S = 0) = \int_{0}^{1} (1+\theta)e^{-t(1+\theta)}\,dt = 1 - e^{-(1+\theta)}$$

Appendix D

Critical values of the Chi-square distribution with d degrees of freedom

d	Probability of exceeding the critical value				
	0.10	0.05	0.025	0.01	0.001
1	2.706	3.841	5.024	6.635	10.828
2	4.605	5.991	7.378	9.210	13.816
3	6.251	7.815	9.348	11.345	16.266
4	7.779	9.488	11.143	13.277	18.467
5	9.236	11.070	12.833	15.086	20.515
6	10.645	12.592	14.449	16.812	22.458
7	12.017	14.067	16.013	18.475	24.322
8	13.362	15.507	17.535	20.090	26.125
9	14.684	16.919	19.023	21.666	27.877
10	15.987	18.307	20.483	23.209	29.588
11	17.275	19.675	21.920	24.725	31.264
12	18.549	21.026	23.337	26.217	32.910
13	19.812	22.362	24.736	27.688	34.528
14	21.064	23.685	26.119	29.141	36.123
15	22.307	24.996	27.488	30.578	37.697
16	23.542	26.296	28.845	32.000	39.252
17	24.769	27.587	30.191	33.409	40.790
18	25.989	28.869	31.526	34.805	42.312
19	27.204	30.144	32.852	36.191	43.820
20	28.412	31.410	34.170	37.566	45.315

Solutions to Odd-Numbered Exercises

Solutions to even-numbered exercises, worked out in full, are available to instructors online in the Instructor's Resource Library.

CHAPTER 1

1.1 The three genotype frequencies are:

$$f_{CC} = 42/(42 + 16 + 32) = 0.467$$
$$f_{CT} = 16/(42 + 16 + 32) = 0.178$$
$$f_{TT} = 32/(42 + 16 + 32) = 0.356$$

The two allele frequencies are:

$$f_C = \frac{2 \times 42 + 16}{2 \times (42 + 16 + 32)} = 0.556$$

$$f_T = \frac{2 \times 32 + 16}{2 \times (42 + 16 + 32)} = 0.444$$

1.3 Let N_{CC}, N_{CT}, and N_{TT} be the observed genotype counts from Exercise 1.1, and let $N = N_{CC} + N_{CT} + N_{TT}$ denote the total sample size. To test the null hypothesis of HWE, we compute

$$\chi^2 = \frac{(N_{CC} - E[N_{CC}])^2}{E[N_{CC}]} + \frac{(N_{CT} - E[N_{CT}])^2}{E[N_{CT}]} + \frac{(N_{TT} - E[N_{TT}])^2}{E[N_{TT}]}$$

$$= \frac{(N_{CC} - Nf_C^2)^2}{Nf_C^2} + \frac{(N_{CT} - 2Nf_C f_T)^2}{2Nf_C f_T} + \frac{(N_{TT} - Nf_T^2)^2}{Nf_T^2}$$

$$= 36.88$$

The critical χ^2 value for $p = 0.05$ and one degree of freedom is 3.841. These data deviate significantly from HWE at the 5% significance level.

1.5 The genotype frequencies are:

$$f_{AA} = f_{AC} = 10/(10 + 10 + 5 + 20 + 5 + 20) = 0.143$$
$$f_{AT} = f_{CT} = 5/(10 + 10 + 5 + 20 + 5 + 20) = 0.0714$$
$$f_{CC} = f_{TT} = 20/(10 + 10 + 5 + 20 + 5 + 20) = 0.286$$

The allele frequencies are:

$$f_A = f_{AA} + (f_{AC} + f_{AT})/2 = 0.250$$
$$f_C = f_{CC} + (f_{AC} + f_{CT})/2 = 0.393$$
$$f_T = f_{TT} + (f_{AT} + f_{CT})/2 = 0.357$$

1.7 To test this triallelic data set for deviation from HWE, we compute

$$\chi^2 = \sum_{a \in \{A,C,T\}} \left(\frac{(N_{aa} - Nf_a^2)^2}{Nf_a^2} + \sum_{b \neq a} \frac{(N_{ab} - 2Nf_a f_b)^2}{2Nf_a f_b} \right) = 45.24$$

The three-allele chi-square test has $3 \cdot (3-1)/2 = 3$ degrees of freedom, so a X^2 value of 45.24 exceeds the $p = 0.05$ critical value of 7.815. Therefore, this data set deviates from HWE at the 5% significance level.

1.9 Consider a locus with two alleles a and A. Let f_A and m_A denote the frequency of allele A in females and males, respectively. After one generation of mating, the genotype frequencies will be the same in males in females:

$$f_{aa}^{(1)} = m_{aa}^{(1)} = m_a f_a$$
$$f_{Aa}^{(1)} = m_{Aa}^{(1)} = m_a f_A + m_A f_a$$
$$f_{AA}^{(1)} = m_{AA}^{(1)} = m_A f_A$$

The allele frequencies will be the following:

$$f_a^{(1)} = m_a^{(1)} = m_a f_a + (m_a f_A + m_A f_a)/2$$
$$f_A^{(1)} = m_A^{(1)} = m_A f_A + (m_a f_A + m_A f_a)/2$$

It is clear that these frequencies do not satisfy HWE—for example, $f_{aa}^{(1)} \neq \left(f_a^{(1)} \right)^2$. A second generation of mating will change the genotype frequencies, but leave the allele frequencies unchanged:

$$f_{aa}^{(2)} = [m_a f_a + (m_a f_A + m_A f_a)/2]^2$$
$$f_{Aa}^{(2)} = 2 \times [m_a f_a + (m_a f_A + m_A f_a)/2] \times [m_A f_A + (m_a f_A + m_A f_a)/2]$$
$$f_{AA}^{(2)} = [m_A f_A + (m_a f_A + m_A f_a)/2]^2$$
$$f_a^{(2)} = f_{aa}^{(2)} + f_{Aa}^{(2)}/2 = [m_a f_a + (m_a f_A + m_A f_a)/2] \times (m_a f_a + m_a f_A + m_A f_a + m_A f_A) = f_a^{(1)}$$
$$f_a^{(2)} = f_{AA}^{(2)} + f_{Aa}^{(2)}/2 = [m_A f_A + (m_a f_A + m_A f_a)/2] \times (m_a f_a + m_a f_A + m_A f_a + m_A f_A) = f_A^{(1)}$$

This time, it is clear that

$f_{aa}^{(2)} = \left(f_a^{(2)}\right)^2$, $f_{Aa}^{(2)} = 2f_a^{(2)}f_A^{(2)}$, and $f_{AA}^{(2)} = \left(f_A^{(2)}\right)^2$, meaning that HWE is satisfied after two generations.

CHAPTER 2

2.1 Nucleotide A must eventually be either fixed or lost in the population, so $\Pr(A \text{ is fixed}) + \Pr(A \text{ is lost}) = 1$. By Equation 2.6, the probability that A becomes fixed is 0.1, so the probability that A is lost is 0.9. Only one allele can become fixed, so the probability that both nucleotides A and C go to fixation is 0. Therefore, $\Pr(A \text{ or } C \text{ are fixed}) = \Pr(A \text{ is fixed}) + \Pr(C \text{ is fixed}) = 0.1 + 0.2 = 0.3$.

2.3 Using Equation 2.5 we find equilibrium allele frequency to be $5 \times 10^{-6} / (5 \times 10^{-6} + 10^{-6}) = 5/6$.

2.5 We first need to find the mutation rate. To do this, we rearrange Equation 2.8 to find

$\mu = d_{AB}/(2t_{AB})$
$= 29 \text{ substitutions}/(2 \times 20 \times 10^6 \text{ years})$
$= 7.25 \times 10^{-7} \text{ substitutions/year}$

We can now estimate the divergence time between species A and C as

$$\hat{t}_{AC} = d_{AC}/(2\mu) = \frac{12 \text{ substitutions}}{2 \times 7.25 \times 10^{-7} \text{substitutions/year}} = 8.3 \times 10^6 \text{years}$$

2.7 The probability that an individual leaves no offspring is the probability that none of the $2N$ offspring in the next generation chose it as a parent.

Each of these $2N$ events is independent and occurs with probability $1 - 1/(2N)$, so
$\Pr(a \text{ chosen individual leaves no offspring}) = \left(1 - \frac{1}{2N}\right)^{2N}$. For a large $2N$, this is approximately $e^{-1} = .37$.

2.9 Individuals in generation $t + 1$ choose their parents uniformly at random from the $2N$ individuals of generation t. No matter which parent the first individual picks, the probability that the second individual picks the same parent is $1/(2N)$. Therefore, the total probability of the two individuals having the same parent in the previous generation is $1/(2N)$.

CHAPTER 3

3.1 Using Equation 3.12, the expected TMRCA is $\sum_{k=2}^{5} \frac{2}{k(k-1)} = 1.6$.

Using Equation 3.13, the expected total tree length is

$$2\sum_{k=1}^{4}\frac{1}{k}=4.1667$$

3.3 Using Equation 3.14, we would expect to see

$$\theta\sum_{k=1}^{4}1/k=4\times20,000\times10^{-5}\times2.083=1.667$$ segregating sites in the

DNA sample.

3.5 There are five segregating sites (positions 1, 2, 4, 9, and 17 of the sequence). The average number of pairwise differences is

$$\frac{3\times1+2\times2+2\times2+3\times1+2\times2}{\binom{4}{2}}=3$$

3.7 The folded SFS from Exercise 3.5 consists of three mutations occurring in a singly haplotype plus two mutations occurring in two haplotypes:

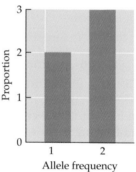

3.9 Let $H_{1,1}$ and $H_{1,2}$ denote the two haplotypes from individual 1, and let $H_{2,1}$, $H_{2,2}$ denote the two haplotypes from individual 2. There are two ways that $H_{1,1}$ and $H_{1,2}$ can be most closely related to each other. The first way is if their ancestral lineages coalesce first. That happens with probability $\binom{4}{2}^{-1}$. The second way is if $H_{2,1}$ and $H_{2,2}$ coalesce first, and then $H_{1,1}$ and $H_{1,2}$ are the next pair to coalesce, which happens with probability $\binom{4}{2}^{-1}\times\binom{3}{2}^{-1}$. The total probability is, therefore,

$$\binom{4}{2}^{-1}\times\left(1+\binom{3}{2}^{-1}\right)=0.222$$

CHAPTER 4

4.1 For population 1, $f_{A1} = (20 + 20/2)/60 = 0.5$. For the second population, $f_{A2} = (15 + 15/2)/60 = 0.375$.

Also,

$$H_T = (f_{A1} + f_{A2})\left(1 - \frac{f_{A1} + f_{A2}}{2}\right) = 0.492$$

$$H_S = f_{A1}(1 - f_{A1}) + f_{A2}(1 - f_{A2}) = 0.484$$

$$F_{ST} = \frac{H_T - H_S}{H_T} = 0.016$$

4.3 a. Using Equation 4.7 we find $E[f_{C1}(t + 1)] = 0.9 \times 0.1 + 0.1 \times 0.9 = 0.18$.

b. After another generation, the expected allele frequency of C in the first population will be $E[f_{C1}(t + 2)] = 0.9 \times E[f_{C1}(t + 1)] + 0.1 \times \{1 - E[f_{C1}(t + 1)]\} = 0.24$.

4.5 a. The expected number of differences between a pair of samples from population 1 is θ_1, because the population size is the same in population 1 and the ancestral population.

b. The effective population size of population 1 is $2N_1 = 10^4$, so the expected number of nucleotide differences is $\theta_1 = 4N_1\mu = 2 \times 10^4 \times 10^{-5} = 0.2$.

For samples from different populations, the expected number of differences is the sum of the expected number of mutations on the lineages in the ancestral population and on the lineages in populations 1 and 2. These are, respectively, θ_A, $\tau\mu$, and $\tau\mu$, where τ is the divergence time in number of generations. The expected number of nucleotide differences is, therefore $\theta_A + 2\tau\mu = 0.2 + 2 \times 6 \times 10^3 \times 10^{-5} = 0.38$.

4.7 From $F_{ST} = T/(T + 2)$, we rearrange to get $T = 2F_{ST}/(1 - F_{ST}) = 00.016$. Converting this to generations by multiplying by $2N = 10^4$ gives 160 generations since the populations split.

4.9 The waiting time to the first migration event is $1/[2M(d - 1)]$ because there are two gene copies, which each migrates into each of $d - 1$ populations at rate M. However, the samples only end up in the same population with probability $1/(d - 1)$, so the equation for $E_D[t]$ becomes

$$E_D[t] = \frac{1}{2M(d-1)} + \frac{1}{d-1}E_S[t] + \frac{d-2}{d-1}E_D[t] \Leftrightarrow E_D[t] = \frac{1}{2M} + E_S[t]$$

Similarly, for samples in the same population, we get

$$E_S[t] = \frac{1}{1 + 2M(d-1)} + \frac{2M(d-1)}{1 + 2M(d-1)}E_D[t]$$

We see that $E_S[t] = d$ and $E_D[t] = d + 1/(2M)$ is the solution to these equations by substitution.

CHAPTER 5

5.1 The frequency of allele A in population 1 is $f_{A1} = (12 + 22/2)/(12 + 22 + 6) = 0.575$. Similarly, the frequency in population 2 is $f_{A2} = (32 + 6/2)/(32 + 6 + 2) = 0.875$. For this pair of populations, $H_S = f_{A1}(1 - f_{A1}) + f_{A2}(1 - f_{A2}) = 0.35375$ and $f_A = (f_{A1} + f_{A2})/2 = 0.725$ $H_T = 2 \times f_A \times (1 - f_A) = 0.39875$. Therefore, $F_{ST} = (H_T - H_S)/H_T = 0.11285$. Based on these data, we can estimate a divergence time of $T = 2F_{ST}/(1 - F_{ST}) = 0.254$ coalescence time units, or $0.254 \times 2N \times T_{gen} = 12{,}721$ years.

5.3 a. Tree A shows evidence of reciprocal monophyly between populations 1 and 2.

b. Tree B best represents the multiregional hypothesis, while tree E best represents the out-of-Africa hypothesis.

c. Trees A, D, E, and F are compatible with the absence of very recent gene flow between populations.

5.5 a. The panda is from population 1 with posterior probability

$$\frac{\Pr(aa \mid 1) \times 0.5}{\Pr(aa \mid 1) \times 0.5 + \Pr(aa \mid 2) \times 0.5} = \frac{f_{1aa}}{f_{1aa} + f_{2aa}} = \frac{6/40}{6/40 + 2/40} = 0.75$$

b. In the case where the panda is estimated to have prior probability 0.9 of being from population 2, its posterior probability of being from population 1 is

$$\frac{\Pr(aa \mid 1) \times 0.1}{\Pr(aa \mid 1) \times 0.1 + \Pr(aa \mid 2) \times 0.9} = \frac{f_{1aa} \times 0.1}{f_{1aa} \times 0.1 + f_{2aa} \times 0.9}$$

$$= \frac{(6/40) \times 0.1}{(6/40) \times 0.1 + (2/40) \times 0.9} = 0.25$$

CHAPTER 6

6.1 There are 51 chromosomes. Let A denote the yellow allele at site 82 and a denote the blue allele, and let B denote the yellow allele at 83 and b denote the blue allele. The haplotype numbers are AB: 22, Ab: 9, aB: 0 and ab: 20. Therefore, $f_{AB} = 0.431, f_{Ab} = 0.176, f_{aB} = 0, f_{ab} = 0.392$, $f_A = 0.607, f_a = 0.392, f_B = 0.431$, and $f_b = 0.568$. $D = 0.431 - (0.607)$ $(0.431) = 0.169$, $D' = 1$ (because there are no aB haplotypes), and $r^2 = (0.169)^2/(0.607 \times 0.392 \times 0.431 \times 0.568) = 0.490$.

6.3 $f_{A_1} = 0.3, f_{A_2} = 0.5, f_{A_3} = 0.2, f_{B_1} = 0.4, f_{B_2} = 0.6$. $D_{11} = 0.12 - (0.3 \times 0.4) = 0, D_{21} = 0.28 - (0.5 \times 0.4) = 0.08, D_{31} = 0 - (0.2 \times 0.4) = -0.08, D_{12} = 0.18$

$- (0.3 \times 0.6) = 0$, $D_{22} = 0.22 - (0.5 \times 0.6) = -0.08$, $D_{32} = 0.2 - (0.2 \times 0.6) = 0.08$. This contrived data set illustrates that if there are more than two alleles per locus, some haplotypes may be in linkage equilibrium and others not.

6.5 a. Each individual in the F_1 population has an AB chromosome from one parent and an ab chromosome from the other. Therefore, $f_{AB} = f_{ab} = f_A = f_B = \frac{1}{2}$ and $D = \frac{1}{2} - (\frac{1}{2})(\frac{1}{2}) = \frac{1}{4}$.

b. Because $c = \frac{1}{2}$, every individual in the F_1 population produces equal numbers of the four possible haplotypes, $f_{AB} = f_{Ab} = f_{aB} = f_{ab} = \frac{1}{4}$. Therefore, in the F_2, $f_A = f_B = \frac{1}{2}$ and $D = 0$.

c. Equation 6.8 seems to predict that D in the F_2 population should be $\frac{1}{8}$ because D should decrease by a factor of $\frac{1}{2}$ each generation if $c = \frac{1}{2}$. Instead, $D = 0$. The reason for the difference is that the F_1 population was not created by random mating. Instead, it was created by hybridizing two homozygous populations. As a consequence, every individual is doubly heterozygous. To derive Equation 6.8, we assumed that the genotype frequencies at each locus are at the Hardy–Weinberg frequencies, which is not true in the F_1.

6.7 a. In population 1, $f_A = 0.7$, $f_B = 0.8$ and $f_{AB} = 0.7$. Therefore, $D = 0.7 - 0.56 = 0.14$. D is the same in population 2.

b. In the mixture, $f_{AB} = f_{ab} = 0.45$ and $f_{Ab} = f_{aB} = 0.05$, so $f_A = f_B = 0.5$. Therefore $D = 0.45 - 0.25 = 0.2$. D is larger by 0.06 because of the population mixture. This is the two-locus Wahlund effect.

c. $D' = 1$ in both populations because each is missing one of the haplotypes. In the mixture, the maximum of D is 0.25, and hence $D' = 0.2/0.25 = 0.8$. The two-locus Wahlund effect will cause D to be larger than the average in the two populations, but D' may be larger or smaller than the average of the values in the two populations.

6.9 a. The χ^2 value is 20.83, which implies $P < 5 \times 10^{-6}$. Whether this result is taken as evidence of a significant association depends on how many tests have been done. If it is the only SNP tested, then there is a highly significant association. But if this is one of one million SNPs tested, you would expect to have five with this P value or smaller.

CHAPTER 7

7.1 The problem tells you that if A is the mutant, $v_A = 1.01v_a$, which implies that $s = 1 - 1/1.01 \approx 0.01$.

a. From the formula in Box 7.1, $t \approx 437$.

b. $t \approx 4370$

7.3 $w_A = 0.95v_a$, which implies that $s = 1 - 1/0.95 = -0.053$. From the formula in Box 7.1, $t \approx 45.2$ generations.

7.5 At equilibrium Equation 7.3 implies that $0.11 = s_{BB}/(s_{BB} + 0.5)$. Therefore, $s_{BB} = 0.062$.

7.7 a. $\bar{v} = 0.8^2 \times 0.85 + 2 \times 0.8 \times 0.2 + 0.2^2 \times 0.1 = 0.868$.

b. Among the newborns, the genotype frequencies are $f_{AA} = 0.64$, $f_{AS} = 0.32$, and $f_{SS} = 0.04$. The relative chances of surviving to reproductive age are 0.85, 1.0, and 0.1. Therefore the genotype frequencies at reproductive age are $f'_{AA} = 0.64 \times 0.85 / \bar{v}$, $f'_{AS} = 0.32 / \bar{v}$, and $f'_{SS} = 0.04 \times 0.1/\bar{v}$, where $\bar{v}$ is calculated as in part a. The answers are $f'_{AA} = 0.626$, $f'_{AS} = 0.369$, $f'_{SS} = 0.005$.

7.9 Here is a table of mating pairs and their offspring:

Mother	Father	Frequency	Fertility	AA	Aa	aa
AA	AA	$\frac{1}{16}$	1	1		
AA	Aa	$\frac{1}{8}$	1	$\frac{1}{2}$	$\frac{1}{2}$	
AA	aa	$\frac{1}{16}$	1		1	
Aa	AA	$\frac{1}{8}$	1	$\frac{1}{2}$	$\frac{1}{2}$	
Aa	Aa	$\frac{1}{4}$	1	$\frac{1}{4}$	$\frac{1}{2}$	$\frac{1}{4}$
Aa	aa	$\frac{1}{8}$	1		$\frac{1}{2}$	$\frac{1}{2}$
aa	AA	$\frac{1}{16}$	1		1	
aa	Aa	$\frac{1}{8}$	1		$\frac{1}{2}$	$\frac{1}{2}$
aa	aa	$\frac{1}{16}$	$\frac{1}{2}$			1

The average fertility is $1 - \frac{1}{32} = \frac{31}{32}$. Among the newborns,

$f_{AA} = [\frac{1}{16} + (\frac{1}{2})(\frac{1}{8}) + (\frac{1}{2})(\frac{1}{8}) + (\frac{1}{4})(\frac{1}{4})](\frac{32}{31}) = \frac{8}{31}$

$f_{Aa} = [(\frac{1}{2})(\frac{1}{8}) + \frac{1}{16} + (\frac{1}{2})(\frac{1}{8}) + (\frac{1}{2})(\frac{1}{4}) + (\frac{1}{2})(\frac{1}{8}) + \frac{1}{16} + (\frac{1}{2})(\frac{1}{8})]$
$(\frac{32}{31}) = \frac{16}{31}$

$f_{aa} = [(\frac{1}{4})(\frac{1}{4}) + (\frac{1}{2})(\frac{1}{8}) + (\frac{1}{2})(\frac{1}{8}) + (\frac{1}{2})(\frac{1}{16})](\frac{32}{31}) = \frac{7}{31}$

b. Among the newborns, $f_A = f_{AA} + f_{Aa}/2 = \frac{16}{31} = 0.5161$. $0.5161^2 = 0.2664 \neq f_{AA} = \frac{8}{31} = 0.2581$. Therefore the genotype frequencies are not in their HW proportions. The difference is small, but it means you cannot assume HW genotype frequencies when working with fertility selection.

7.11 The table analogous to Table 7.1 if R is recessive is:

Father	Mother	Frequency	Offspring viability
RR	Rr	$f_{RR}f_{Rr}$	$1-s/2$
Rr	Rr	f_{Rr}^2	$1-s/4$
RR, Rr, rr	RR	f_{RR}	1
RR, Rr, rr	rr	f_{rr}	1
rr	Rr	$f_{rr}f_{Rr}$	1

Half of the offspring of the $RR \times Rr$ pairs will be Rh$^+$, but only $\frac{1}{4}$ of the offspring of the $Rr \times Rr$ will be Rh$^+$, so the average viability loss in the first type of family will be twice the loss in the second type of family. In all other families, either the mother will be Rh$^+$ (when the mother is RR) or the offspring will be Rh$^-$ (when either parent is rr).

b. Yes. The same logic used for the case with R dominant shows that both R and r will decrease in frequency when rare.

7.13

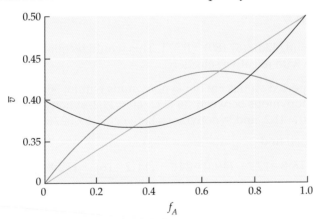

CHAPTER 8

8.1 The general formula is $r = 2N\mu u(s,N)$ where u is given by Equation 8.1. Note that s has to be negative in the formula for u.

a. N = 10,000, $r = 3.7 \times 10^{-25}$; N = 1000, $r = 1.6 \times 10^{-10}$; $r = 1.8 \times 10^{-9}$.

b. N = 10,000, $r/\mu = 1.7 \times 10^{-16}$; $r/\mu = 0.075$; $r/\mu = 0.81$.

8.3 Solve $0.002 = \sqrt{2s/(10,000\pi)}$ for s. $s \approx 0.063$. Selection has to be much stronger on recessive advantageous alleles than on advantageous alleles with additive effect to have the same fixation probability.

8.5 For insulin: $1 - \alpha = 0.13/2.2 = 0.06$ so $\alpha = 0.94$. For histones: $1 - \alpha = 10^{-4}/2.2 = 4.5 \times 10^{-5}$ so $\alpha = 0.999955$.

8.7 The total substitution rate, r, is the sum of the rates for advantageous alleles and for neutral alleles. A fraction 0.0008 of the sites are advantageous and their substitution rate is $(2N\mu)(2s) = 4Ns\mu$. A fraction $1 - \alpha - 0.0008$ is neutral, and its substitution rate is μ. Therefore the net rate is $(0.0008 \times 4 \times 10,000 \times 0.01 + 1 - \alpha - 0.0008)2.2 \times 10^{-9} = 0.8 \times 10^{-9}$. Solve for α to find $\alpha = 0.956$.

8.9 $c \approx \ln[(\delta - f_B)/(1 - f_B)]/t$ with $t = 100, f_B = 0.03$ and $\delta = 139/146 = 0.952$. Therefore $c \approx 0.00051 = 0.05$ cM. If 1 cM = 1 mb, the causative gene would be about 50 kb from *CSF1R*. A more sophisticated version of this result guided researchers to finding the causative gene at about 70 kb from *CSF1R*.

CHAPTER 9

9.1 $d_N/d_S = (8/420)/(6/180) = 0.571$. This would be qualitatively compatible with negative selection.

9.3 The expected values are nonsynonymous within: $36 \times 28/60 = 16.8$, nonsynonymous between: $36 \times 32/60 = 19.2$, synonymous within: $24 \times 28/60 = 11.2$, and synonymous between: $24 \times 32/60 = 12.8$.

We then find

$$X^2 = \frac{(16.8-12)^2}{16.8} + \frac{(19.2-24)^2}{19.2} + \frac{(11.2-16)^2}{11.2} + \frac{(12.8-8)^2}{12.8} = 6.429$$

As this value is larger than the critical value of 3.841 at the 5% significance level, we reject the null hypothesis of an equal ratio of nonsynonymous and synonymous mutations within and between species. The ratio of nonsynonymous to synonymous mutations is highest between species, and is compatible with the hypothesis of positive selection.

CHAPTER 10

10.1 The average contribution of a nonmutant individual to the next generation is $2/N$ when the mutant is in very low frequency. The average contribution of a mutant individual is $\dfrac{2}{N} + \dfrac{\delta m}{N}\left(\dfrac{1}{m} - \dfrac{1}{1-m}\right)$.

The ratio is $1+\dfrac{\delta m}{2}\left(\dfrac{1}{m}+\dfrac{1}{1-m}\right)$, which is approximately 1.041 if $m = 0.4$ and $\delta m = 0.05$. Therefore $s = 0.041$.

10.3 a. Substituting into Equation 10.5 with $b = 10$, $c = 8$ and $d = 5$, $\hat{f} = {}^{3}\!/_{13}$ is the ESS frequency of fighting.

 b. You know that f is at a value that makes the coefficient of δf in Equation 10.3 equal to 0, because that is the condition used to obtain Equations 10.4 and 10.5. The problem tells you that d is decreased somewhat because of the smaller number of females. That is, d is changed to $d' = d - \delta d$. In Equation 10.3, that changes the coefficient of δf from 0 to $-(1 - f)(- \delta d)$, which is positive. Therefore, any mutation that increases f will increase in frequency. Another way to reach the same conclusion is to ask whether the ESS value of f is larger or smaller when d is decreased slightly. The derivative of $\hat{f}$ as a function of d is negative; hence, reducing d will increase $\hat{f}$.

10.5 Haldane recognized that each brother shared half of his genes on average, so saving two brothers would be equivalent to saving himself. In a similar way, each cousin shares on average ⅛ of his genes, so saving eight cousins would be equivalent to saving himself.

10.7 a. Among the males, ½ will carry D and half will carry d. The d-bearing males will produce equal numbers of X and Y gametes, so their offspring will be in a 1:1 sex ratio. The D-bearing males will produce only X-bearing gametes so all their offspring will be female. If D-bearing and d-bearing males are equally successful in mating, then the sex ratio in the next generation will be ¾ female and ¼ males.

 b.

Family	Frequency	$X^D X^D$	$X^D X^d$	$X^d X^d$	$X^D Y$	$X^d Y$
$X^D X^D \times X^D Y$	⅛	1				
$X^D X^d \times X^D Y$	¼	½	½			
$X^d X^d \times X^D Y$	⅛		1			
$X^D X^D \times X^d Y$	⅛		½		½	
$X^D X^d \times X^d Y$	¼		¼	¼	¼	¼
$X^d X^d \times X^d Y$	⅛			½		½
Total		¼	⅜	⅛	⅛	⅛

The totals are obtained by multiplying the frequency of each family by the outcome of the mating. This table confirms the answer to part a: $3/4$ of the offspring will be female and $1/4$ will be male. Among the males, half will carry D and the other half will carry d. Among the females, $1/3$ will be DD, $1/2$ will be Dd, and $1/6$ will be dd, which implies the frequency of D in females is $7/12$. To compute the overall average frequency of D on X chromosomes, we need to remember that females carry two X's and males only 1. Since $3/4$ of the population is female, $6/7$ of the X's are in females and $1/7$ are in males. Therefore the overall frequency of D is $(6/7)(7/12) + (1/7)(1/2) = 4/7$.

c. D will increase in frequency every generation, which means that the proportion of males will decrease every generation until the population goes extinct because there are too few males. Usually, what happens is that other mutations at other loci arise that suppress the effect of the D allele, but a distorter allele may, in principal, drive a population to extinction.

10.9 The change in f_A because of meiotic drive is $rf_A f_a$ and the decrease because of selection is $-sf^2_A f_a$. These two terms are equal when $f_A = r/s = 0.01/0.05 = 1/5$.

10.11 The genotype frequencies among zygotes are 0.0001, 0.0198 and 0.9801. The average viability is $w = 0.0001 + 0.0099 + 0.49005 = 0.50005$. Among the adults, the frequencies are $0.0001/0.50005 \approx 0.0002$, $0.0099/0.50005 \approx 0.0198$ and $0.49005/0.50005 \approx 0.98$. The effect of immigration is given by Equation 9.14 with $m = 0.1$: $f_{AA} \approx 0.00018$, $f_{Aa} \approx 0.01782$, $f_{aa} \approx 0.982$. In the next generation, then, $f_A = f_{AA} + f_{Aa}/2 = 0.0091$. The frequency of A has decreased because selection is inefficient when A is in low frequency.

CHAPTER 11

11.1 Using the formulas, $\bar{x} = \dfrac{1}{n}\sum_{1}^{n} x_i$, $V = \dfrac{1}{n}\sum_{i=1}^{n}(x_i - \bar{x})^2 = \dfrac{1}{n}\sum_{i=1}^{n} x_i^2 - \bar{x}^2$,

and $\sigma = \sqrt{V}$, $\bar{x} = 88.3$ g, $V = 23.8$ g^2, and $\sigma = 4.9$ g. If your calculator returned 26.5 for V and 5.1 for σ, it used the formula used the formula $V = \dfrac{1}{n-1}\sum_{i=1}^{n}(x_i - \bar{x})^2$ instead. That is not incorrect. For some purposes the $n - 1$ is better than n. Dividing by n gives unbiased estimates of the variance and covariance while dividing by n gives the maximum likelihood estimates.

11.3 a. The mean of the parents is 656.4 g and the mean of the selected parents is 720.6 g. Therefore $S = 64.2$ g. $h^2 = 0.64$. Therefore, $R = 0.64 \times 64.2 = 41.1$. The mean of the offspring should be $656.4 + 41.1 = 697.5$ g.

   b. Now the mean of the selected parents is 778 so $S = 121.6$, $R = 77.8$ and the mean offspring weight should be $656.4 + 77.8 = 734.2$ g.

   c. The problem with selecting only a few parents is that the population size is so small that deleterious alleles can increase in frequency and become fixed by genetic drift, with the result that overall viability will be reduced.

11.5 The breeding values for each genotype are given in Table 11.4. The average breeding value is obtained by multiplying the breeding value for each genotype by the Hardy–Weinberg frequency for that genotype:

$$f_A^2[2f_a\alpha] + 2f_Af_a[-\alpha(f_A - f_a)] + f_a^2[-2\alpha f_A] = 0$$

11.7 Substituting in Equation 11.13, $V_A = 2\alpha f_A^2(1 - f_A)$ and $V_D = 4f_A^2$ $(1 - f_A)^2\alpha^2$. The graphs assume $\alpha = 1$.

Father	Mother	Frequency	Offspring viability
RR	Rr	$f_{RR}f_{Rr}$	$1-s/2$
Rr	Rr	f_{Rr}^2	$1-s/4$
RR, Rr, rr	RR	f_{RR}	1
RR, Rr, rr	rr	f_{rr}	1
rr	Rr	$f_{rr}f_{Rr}$	1

11.9 $\bar{x} = x_{aabb} + f_A^2 f_B^2 \alpha = 10 + 0.7^2 0.8^2 = 10.3136$ and

$$V_G = \left(1 - f_A^2 f_B^2\right)\left(x_{aabb} - \bar{x}\right)^2 + f_A^2 f_B^2 \left(x_{aabb} + \alpha - \bar{x}\right)^2 = 0.215...$$

Glossary

A

additive genetic variance, V_A The variance in breeding values in a population; one component of the total genetic variance.

additive model A model of a quantitative character in which the deviations from a reference genotype are added. For a single locus, the additive model assumes there is no dominance. For more than one locus, the additive model assumes there are no epistatic interactions among loci.

admixed A population is said to be admixed if it has received gene-flow from another population. Usually only recent gene flow is considered. The concept is sometimes also used to describe individuals that have ancestors from several different populations.

ancestral allele The allele existing before a mutation occurs.

ancestral lineages Here, used for the branches (or edges) of a coalescence tree. More generally, the term refers to connected paths of descendants through a genealogy.

Approximate Bayesian Computation (ABC) A simulation-based technique for approximating posterior probabilities.

assortative mating A mating structure in which pairs of individuals that are (genetically) similar to each other mate with higher probability than expected under random mating.

asymptotic efficiency A estimator is asymptotically efficient if it attains the minimal possible variance as the sample size goes to infinity.

average number of pairwise differences The number of pairwise differences between two sequences is the number of positions in the DNA in which the two sequences differ. The average number of pairwise differences for a sample of sequences is found by averaging the number of pairwise differences over all pairs of sequences in the sample.

B

Bayes' law Bayes' law states, in its simplest form, that for two events A, B, assuming

$$\Pr(B) > 0: \ \Pr(A\,|\,B) = \frac{\Pr(B\,|\,A)\Pr(A)}{\Pr(B)}$$

Bayesian methods Methods that combine a likelihood function with a prior probability to calculate a posterior probability. The posterior probability can be used to estimate parameters and to quantify available knowledge regarding the parameter.

Bayesian statistics See Bayesian methods.

Bernoulli RV A discrete random variable describing the outcome of a coin toss of a possibly biased coin (see Appendix A, p. 235).

binomial RV A discrete random variable describing the number of successes when a fixed number of trials with constant success probability has been carried out (see Appendix A, p. 234).

biometrical analysis The study of quantitative characters using only phenotypic measurements including means, variances, and covariances.

biometry The same as biometrical analysis. Also, statistics that are commonly applied to biological data.

breeder's equation $R = h^2S$, where R is the change in the population mean in one generation of selection and S is the selection differential.

breeding value   The net effect of genetic factors that are transmitted on average to offspring.

broad-sense heritability, h_B^2 **(also denoted by** H^2**)**   The ratio of the total genetic variance (V_G) of a quantitative character to the total variance (V_x).

C

categorical data   Data in the form of numbers of observations from each of a fixed number of discrete categories.

coalesced   Two lineages have coalesced in a coalescence tree when, looking backwards in time, the have found a most recent common ancestor (a coalescent event has occurred).

coalescence event   The coalescence (merging) of two ancestral lineages in a coalescence tree, occurring at the time a set of individuals last had a common ancestor.

coalescence process   A stochastic process that models the ancestry of a sample. The outcome of the coalescence process is a coalescence tree.

coalescence theory   A theory describing the ancestry of a sample in terms of a coalescence tree.

coalescence time   The time at which a coalescence event occurs.

coalescence tree   A tree representing the ancestry of a sample in which edges (also called branches or lineages) represent lines of descent and nodes represent coalescence events.

coalescent   Synonymous with coalescence event.

coalescent effective population size   The number of individuals in a standard coalescence model needed to generate the same rate of coalescence between pairs of lineages as that observed (or inferred) for the real population.

coefficient of linkage disequilibrium (D)   The difference between the frequency of a haplotype in a population and the product of the allele frequencies (see Equation 6.1).

common ancestor   An ancestor shared by two or more individuals.

conditional probability   The conditional probability of the event A, given the event E has occurred, is defined as $\Pr(A \mid E) = \Pr(A$ and $E)/\Pr(E)$.

confidence interval   An $x\%$ confidence interval for a parameter is an interval that includes the true value of the parameter with probability $x\%$. It is used as a measure of statistical confidence (certainty) in classical statistics.

consistency   An estimator is a consistent estimator of a parameter if it converges to the true value of the parameter as the sample size increases.

continuous random variable   A random variable defined on a continuous sample space.

correction for multiple hits   Estimation of distances between DNA sequences that are expected to be proportional to time, and that take into account that each site could be hit by more than one mutation in the history of the sequences.

credible intervals   A credible interval is a concept used in Bayesian statistics to quantify statistical uncertainty. The interval (a, b) is a $C\%$ credible interval for the scalar parameter θ if $\Pr(a < \theta < b \mid X) = C/100$, where X is the observed data.

cumulative distribution function (CDF)   The cumulative distribution function (CDF) of a random variable (X) is defined as $F \times (x) - \Pr(X \leq x)$.

D

deme   A subpopulation, or subset of a larger population. Often used to describe the smallest unit of individuals that can be described as evolving with random mating.

derived allele   The allele generated by a new mutation.

di-allelic model   A model which assumes that at most two alleles segregate in the population.

directional selection   Selection that favors one allele over another at the same locus regardless of the allele frequency.

dis-assortative mating   A mating structure in which pairs of individuals that are (genetically) dissimilar to each other mate with higher probability than expected under random mating.

disruptive selection   Selection that favors a rare genotype.

distance-based methods   In phylogenetic inference, methods that estimate the tree topology by first estimating a distance matrix for all pairs of DNA sequences and then identify a tree, or set of trees, that fit the distance matrix according to an algorithmic criterion.

divergence model   A model that describes the history of a set of populations in terms of splitting (divergence) between populations, but without inclusion of gene-flow.

DNA fingerprinting A technique, often used in forensics, that uses DNA to test identity between a sample and an individual.

DNA profiling Synonymous with DNA fingerprinting. A DNA profile is the combination of genotypes (the DNA fingerprint) in an individual used in DNA fingerprinting.

dominance deviation, δ The difference between the mean of a character in heterozygous individuals and the average of the means of the two classes of homozygous individuals, i.e., $\delta = [x_{Aa} - (x_{AA} + x_{aa})/2]$.

E

ecological genetics The study of survival and reproduction of individuals with different genotypes and phenotypes in natural populations of plants and animals.

edge Here used synonymously with a branch in a tree or a linage in a tree. *Edge* is a term borrowed from graph theory, where it indicates the connection between two nodes in a graph.

effective population size The effective population size is the number of individuals in an idealized population (such as a Wright–Fisher population) that would generate the same value of a defined statistic (or other property of the population) as that observed for the real population. Many different statistics are used, but heterozygosity is possibly the most commonly used for defining effective population sizes.

environmental variance, V_E The variance of a quantitative character among individuals with the same genotype.

epistasis In quantitative genetics, there is epistasis if the genetic effects of different loci cannot be added. In transmission genetics, there is epistasis if the phenotypes associated with a gene cannot be seen unless another gene is expressed.

estimate A statistical guess of the true value of a parameter.

estimator A statistical procedure for generating an estimate.

evolutionarily stable A condition in a population that will resist changes resulting from mutations of small effect. Mutations that change the condition will not tend to increase in frequency.

evolutionarily stable strategy (ESS) A phenotype or behavior that is evolutionarily stable.

evolutionarily unstable A condition in a population that will not persist if mutations arise that modify it slightly. Those mutations will tend to increase in frequency.

Ewens sampling formula This formula provides the probability of obtaining a particular sample configuration under several different models, including the standard coalescence model, when mutations occur according to an infinite alleles model.

expected heterozygosity The proportion of heterozygous individuals expected under a specific population genetic model, typically under the assumption of random mating.

expected homozygosity The proportion of homozygous individuals expected under a specific population genetic model, typically under the assumption of random mating.

exponential distribution A continuous distribution often used to model the waiting time until the first event, when events occur at a constant rate (λ). The exponential distribution has sample space of $(0, \infty)$ and probability density function $f(x) = \lambda \exp(-\lambda x)$, $\lambda > 0$ (see also Appendix B).

external lineages Lineages (edges) in a tree connected to leaf nodes.

F

F-statistics Statistics measuring reductions in heterozygosity. They form the basis of a body of population genetic theory developed by S. Wright. Examples of F-statistics include the inbreeding coefficient (denoted F in this book) and F_{ST}.

Felsenstein's equation This equation is given by $\Pr(X | \Theta) = \int_G \Pr(X | G)p(G | \Theta)dG$, where X is the data (typically DNA sequence data from multiple individuals from one or more populations), G is the coalescence tree defined in terms of topology and branch lengths, and Θ is a set of population genetic parameters of interest. The integral is over the set of all possible coalescence trees.

fitness Fitness can be defined differently in different models. For a diploid genotype it is typically defined as the expected number of offspring of an individual of that genotype left to reproduce in the next generation.

folded frequency spectrum A representation of counts of sample allele frequency observations for a set of DNA sequence from multiple individuals that does not require knowledge of which allele is ancestral and which is derived for each SNP. A folded site frequency spectrum is obtained from an unfolded spectrum, i.e., a $n-1$ dimensional vector for n sequences $\mathbf{f} = (f_1, f_2, \ldots, f_n)$, as

$$f_i^* = \begin{cases} f_i + f_{n-i} & \text{if } i < [n/2] \\ f_i & \text{if } i = [n/2] \end{cases}, i = 1, 2, \ldots, [n/2],$$

where $[n/2]$ is the value of n rounded down to nearest integer value.

founder effect The effect of a temporary decline in population size on genetic variation as a new population, or species, is founded.

fourfold degenerate site A site at the third position in a codon for an amino acid at which any of the four nucleotides (A, T, G, C) results in the same amino acid.

G

gametic self-incompatibility A mechanism in some plant groups that prevents the fertilization of flowers by pollen carrying particular alleles at a self-incompatibility locus.

gene-flow Exchange of alleles between (sub) populations due to migration.

genetic drift A change in allele frequencies over time in a population of finite size due to random transmission of parental alleles from parents to offspring and due to the fact that some individuals randomly (irrespective of genotype) produce more offspring than other individuals.

genic selection Selection that occurs because each copy of an allele affects viability independently.

genome-wide association study (GWAS) A study that is designed to find a nonrandom association between marker loci spread throughout the genome and a disease or a phenotypic trait.

genotype The combination of alleles found in an individual at a particular locus.

genotype–environment interaction (G×E) The dependence of the average phenotype produced by a given genotype on the environment experienced during growth and development.

geometric RV The geometric random variable is often used to describe the number of successes before the first failure in repeated independent

trials with constant success probability (see Appendix A, p. 241).

group selection Selection that results from the overall survival and reproduction of a group of individuals.

H

Hamilton's Rule The rule that an allele causing altruistic behaviors will increase in frequency if the cost (c) to the individual performing the behavior is less than the gain to close relatives (b) multiplied by the coefficient of relatedness (R), $c < Rb$.

haplotype The combination of alleles at two or more loci on a chromosome.

harmonic mean The harmonic mean of k numbers, $x_1, x_2, \ldots x_k$, is given by

$$\frac{k}{\frac{1}{x_1} + \frac{1}{x_2} + \ldots + \frac{1}{x_k}}.$$

heterozygosity The proportion of individuals in a population that are heterozygous at a particular locus.

heterozygote advantage Selection that favors individuals heterozygous at a given locus.

heterozygote disadvantage Selection that favors individuals homozygous at a given locus.

HKA test A test of neutrality based on comparing variability within and between species in multiple loci.

homozygosity The proportion of individuals in a population that are homozygous at a particular locus.

horizontal gene transfer Gene-flow between groups that have been defined as different species.

I

identical by descent (IBD) Two alleles are identical by descent if the are identical because of shared ancestral descent (in contrast to identity caused by two identical mutations).

inbreeding What happens when individuals related to each other produce offspring together.

inbreeding coefficient The inbreeding coefficient measures the excess of homozygous individuals in a population relative to the expectation under Hardy-Weinberg Equilibrium (see p. 13).

incomplete lineage sorting   Absence of reciprocal monophyly due to preservation of shared ancestral variation.

independence   See independent.

independent   Two random variables X and Y are independent if, and only if, $\Pr(X = x \text{ and } Y = y) = \Pr(X = x)\Pr(Y = y)$.

infinite alleles model   A model of mutation that assumes all new mutations generate a new allele, i.e., there is a countably infinite number of alleles and no back-mutation.

infinite sites model   A model of mutation that assumes the same site in a sequence never can be hit by more than one mutation, i.e., each mutation generates a new segregating site.

interaction variance, V_I   The part of the total genetic variance that is not attributable to additive effects. The interaction variance includes both the dominance and epistatic variances.

internal lineages   Lineages (edges) that connect two internal nodes.

internal nodes   Nodes in the tree that are not leaf nodes. In a coalescence tree, internal nodes represent most recent common ancestors of subsets of the sample.

island model   A model of population structure with a fixed number of demes, with random mating within demes, and possibly exchange of migrants between demes.

isolation by distance   There is isolation by distance among a set of populations if genetic divergence (for example measured using F_{ST}) is correlated with geographic divergence.

J

joint frequency spectrum   The frequency spectrum for two or more populations considered together. While the frequency spectrum for one population based on a sample of size n is a vector of $n + 1$ entries when including invariable sites, the joint frequency spectrum for two populations with sample sizes n_1 and n_2, when including invariable sites, is a vector of dimension $(n_1 + 1) \times (n_2 + 1)$.

joint probability   The probability distribution for two or more random variables.

K

k-allelic locus   A locus in which there are k different alleles, where k could be any positive natural number.

kin selection   Selection that results from the effect of an allele on an individual and on close relatives whose survival and reproduction are affected by the individual.

L

law of total probability   The law of total probability states that if $A_1, A_2, \ldots A_r$, are mutually exclusive events and $\sum_{i=1}^{r} \Pr(A_i) = 1$,

then for any event E $\Pr(E) = \sum_{i=1}^{r} \Pr(E \mid A_i)\Pr(A_i)$.

laws (axioms) of probability   Basic laws of probability assumed to be true, and from which all other theorems regarding probability can be derived (see also Appendix A, p. 233).

leaf   Synonymous with leaf node.

leaf node   A "tip" of a tree, i.e., a node connected to only one other node. The set of leaf nodes of a coalescence tree represents the sample.

likelihood function   The likelihood function is any function proportional to the probability of the data. It is considered a function of parameters of a statistical model. In phylogenetics, one of the parameters is the topology of the phylogenetic tree.

LINEs (Long Interspersed Nuclear Elements)   A type of transposon found in high abundance in the genomes of humans and other mammals.

linkage disequilibrium (LD)   The nonrandom association of alleles at two or more loci on a chromosome.

linkage equilibrium   A condition in which the haplotype frequency in a population is equal to the product of allele frequencies.

M

MacDonald–Kreitman (MK) test   A test of neutrality based on comparing the ratio of nonsynonymous to synonymous mutations within and between species.

Markov Chain Monte Carlo (MCMC)   A simulation-based technique often used to handle problems with missing data or latent variables in statistics. In populations genetics the technique is often used to estimate parameters while taking into account uncertainty regarding the (unknown) structure of the coalescence tree.

match probability   The probability of random identity between two DNA profiles.

maximum likelihood principle   The principle that estimates of parameters based on a particular experiment should be obtained by choosing the value of the parameter(s) that maximize the probability of observing the particular outcome of the experiment.

maximum parsimony method   In phylogenetic inference, a method that estimates tree topology by choosing the tree that requires the fewest mutations.

meiotic drive   The tendency of an allele or haplotype to be overrepresented in the gametes produced by an individual, thus violating Mendel's first law.

migration rates   The migration rate from population i to population j is the proportion of individuals in population j that are replaced by individuals from population i each generation.

molecular clock   Substitutions obey a molecular clock if they occur at a constant rate in time. In phylogenetics, used to define rooted trees in which the sum of the branch-lengths in the path in the tree from the root to a leaf is the same for all leaf nodes.

most recent common ancestor   The most recent common ancestor for a set of individuals is the last individual alive who was an ancestor of all individuals in the set.

multiregional hypothesis   A hypothesis that posits that modern humans evolved simultaneously in many regions of the world.

mutation rate   The average number of new mutations per generation.

N

narrow-sense heritability, h^2   The ratio of the additive genetic variance to the total variance of a quantitative character (V_A/V_x)

negative selection   Selection acting against new mutations, i.e., selection in which the new mutation is associated with a negative selection coefficient.

neighbor-joining algorithm   An algorithm specifying a distance-based method that will give the correct tree if the distance matrix is estimated with no uncertainty, and that does not rely on the assumption of a molecular clock.

neutral theory   In this book, synonymous with the neutral theory of molecular evolution.

neutral theory of molecular evolution   A hypothesis proposed by M. Kimura which posits that molecular variation within and between species can be explained by mutation and genetic drift.

nonsynonymous mutation   A mutation of a single nucleotide in a coding sequence that changes the amino acid coded for.

O

outgroups   A set of species that do not share a most recent ancestor with any members of the reference set (the ingroups) before the ingroups share a most recent common ancestor with each other.

P

panmixia   Synonymous with random mating.

parameter   In statistics, any part of a model that can be estimated from the data.

partial sweep   The change in allele frequencies at neutral sites closely linked to a site carrying an allele that has increased in frequency because of natural selection but that has not yet been fixed.

Poisson RV   A random variable ($\leq \lambda$) often used to describe the number of events occurring in a time interval when events occur at a constant rate (see Appendix A, p. 240).

population structure   There is population structure when matings are more likely to occur between some subsets of the population than between others, typically due to geographic structure; individuals located in geographical proximity to each other are more likely to mate. Population structure is also used to describe a population in which allele frequencies differ between different geographic regions.

population subdivision   Synonymous with population structure.

positive selection   Selection acting in favor of new mutations, i.e., selection in which the new mutation is associated with a positive selection coefficient.

posterior distribution   A distribution of posterior probabilities. A posterior probability represents knowledge regarding a parameter incorporating information gained by considering the data and incorporating information from a prior distribution.

Principal Component Analysis (PCA)   A statistical technique that reduces a data set with many variables into a set of (possibly fewer) uncorrelated variables. It is used in genetics to identify structures in large data sets useful for defining genetic relationships between individuals. The results of PCA analyses in genetics are often presented in terms of a two-dimensional plot where the distances between individuals in the plot summarize some components of the genetic differentiation between the individuals. Individuals closer to each other in the plot are, by some measure of genetic distance, genetically more similar than individuals distant from each other in the plot.

prior distribution   A distribution of prior probabilities. A prior probability represents knowledge of a parameter before data are considered.

probability   The probability of an event can be defined as the proportion of time this event is expected to occur in a long series of repeated experiments.

probability density function (PDF)   The PDF of a continuous RV defines the relative probability of the random variable to take on a particular value. The probability that the RV falls in a particular region is calculated by integrating the PDF over this region.

probability mass function (PMF)   The PMF of a discrete random variable defines the probabilities assigned to each possible event.

Q

quantitative characters   A phenotypic character that is measured on a continuous scale and does not have a simple Mendelian basis.

quantitative trait locus (QTL)   A locus at which different genotypes for a quantitative character have different mean values.

R

random mating   Random mating occurs when all individuals are equally likely to mate with each other irrespective of genotype, phenotype, geographic location, family relationship, etc. For species with two sexes, random mating occurs when all females are equally likely to mate with all males, and vice versa.

random variable   A variable that takes on different values (e.g., possible outcomes of an experiment) and for which each value can be associated with a probability.

rate of substitution   The average number of mutations that goes to fixation (reaches a population frequency of 1) per generation.

realized heritability   The heritability estimated by measuring the response to selection on a

quantitative character. The change per generation in the mean (R) and the selection differential (S) are measured, and the realized heritability is R/S. In practice, R and S are estimated by averaging over several generations of selection.

reciprocal monophyly   A tree is reciprocally monophyletic for two groups if all members of each group share a most recent common ancestor with each other before they share a common ancestor with any members of the other group.

recurrent mutation   The creation of a given allele by mutation more than once.

reinforcement   The process by which assortative mating with nonimmigrant individuals becomes more extreme because of natural selection favoring nonimmigrant individuals. Reinforcement may eventually lead to reproductive isolation.

retrotransposon   A type of transposon derived from a retrovirus. Retrotransposons insert new copies of themselves by first creating an RNA copy and then creating a new DNA copy by reverse transcription.

root   The root of a rooted tree is the only node that connects to exactly two other nodes. In a coalescence tree it represents the most recent common ancestor of the entire sample.

S

sample space   The possible values of a particular random variable.

segregating sites   Positions in a DNA sequence that differ between two or more individuals.

selection differential, S   The difference between the mean of a character after selection ($\bar{x}'$) and the mean before selection ($\bar{x}$).

selective sweep   The change in allele frequencies at neutral sites closely linked to a site carrying an allele that has been driven to fixation by natural selection.

significance level   A concept used in statistical hypothesis testing to determine when to reject a null hypothesis. If the probability of observing an outcome as extreme or more extreme than the observed outcome under the null hypothesis is less than the significance level, then the null hypothesis is rejected. A significance level of 0.05 or 0.01 is chosen in many studies.

Single Nucleotide Polymorphisms (SNPs)   Synonymous with segregating sites.

singletons   Mutations segregating at a frequency of $1/n$ in a sample of n sequences.

site frequency spectrum (SFS)   The counts of sample allele frequency observations for a set of DNA sequence from multiple individuals. For a sample of n sequences it is a vector of length $n-1$ in which the ith element is the number (or sometimes proportion) of SNPs for which the mutant (derived) allele segregates at a frequency of i/n in the sample. This is also known as the "unfolded site frequency spectrum." Versions also exist that include the fixed 0 and n classes of sites.

stable polymorphism   A polymorphism that persists for a long time because it is maintained by a balance of opposing forces.

star phylogeny   A phylogeny in which all leaf nodes are connected to the root.

statistic   A statistic is anything that can be calculated from the data.

stepping-stone models   A set of population genetic models that, in their simplest form, assume a linearly organized number of populations in which migration occurs only between adjacent populations.

sufficiency   An statistic is sufficient for a parameter if it contains all relevant information regarding the parameter. Formally, for data X, the statistic $T(X)$ is sufficient for the parameter θ if $\Pr(X = x \mid T(X) = t, \theta) = \Pr(X = x \mid T(X) = t)$.

synonymous mutation   A mutation of a single nucleotide in a coding sequence that does not change the amino acid coded for.

T

Tajima's D test   A test of neutrality based on a summary of the site frequency spectrum.

Tajima's estimator   An estimator, named after F. Tajima, of the population genetic parameter θ. It is given by the average number of pairwise differences.

total tree length   The sum of the lengths of all lineages (edges, branches) in a tree.

trajectory   The sequence of allele frequencies from some initial to some final time.

transition mutation   A mutation that replaces one purine (A, G) with the other or one pyrimidine (G, C) with the other.

transposon   A short chromosomal region that is capable either of excising itself and inserting itself into another genomic location or of duplicating itself and inserting a copy into another genomic location while remaining at the initial location.

transversion mutation   A mutation that replaces a purine with a pyrimidine or a pyrimidine with a purine.

two-locus Wahlund effect   The creation of linkage disequilibrium when samples from two or more populations are mixed.

V

variance   The variance of a random variable (X) is defined as $V[X] = E[(X - E[X])^2]$.

viability selection   Selection that occurs because individuals with different genotypes differ in their rate of survival from the zygote stage to adulthood.

W

Wahlund effect   The increase in the proportion of homozygotes in a population due to population subdivision.

Wright–Fisher model   The most common population genetic model used to predict the transmission of gene copies between generations. The are many extensions to this model, named after S. Wright and R. A. Fisher, but the standard version assumes a haploid population, here assumed to be of size $2N$, in which the probability that individual j in generation $t+1$ is a descendent of individual i in generation t is $1/(2N)$ for all $i, j = 1, 2, \ldots 2N$, independently for all j.

Photo Credits

Index

Page numbers followed by *b* denote boxes; those followed by *f* denote figures; and those followed by *t* denote tables.

About the Book

Editor: Andrew Sinauer
Project Editor: Martha Lorantos
Copy Editor: Carrie Crompton
Indexer: Sharon Hughes
Production Manager: Christopher Small
Photo Researcher: David McIntyre
Book Design and Layout: Janice Holabird
Illustration Program: Joanne Delphia